AF737472

NEW INTELLECTUAL PROPERTY ISSUES

NEW INTELLECTUAL PROPERTY ISSUES

JOHN V. LUONG
EDITOR

Nova Science Publishers, Inc.
New York

Copyright © 2007 by Nova Science Publishers, Inc.

All rights reserved. No part of this book may be reproduced, stored in a retrieval system or transmitted in any form or by any means: electronic, electrostatic, magnetic, tape, mechanical photocopying, recording or otherwise without the written permission of the Publisher.

For permission to use material from this book please contact us:
Telephone 631-231-7269; Fax 631-231-8175
Web Site: http://www.novapublishers.com

NOTICE TO THE READER

The Publisher has taken reasonable care in the preparation of this book, but makes no expressed or implied warranty of any kind and assumes no responsibility for any errors or omissions. No liability is assumed for incidental or consequential damages in connection with or arising out of information contained in this book. The Publisher shall not be liable for any special, consequential, or exemplary damages resulting, in whole or in part, from the readers' use of, or reliance upon, this material.

Independent verification should be sought for any data, advice or recommendations contained in this book. In addition, no responsibility is assumed by the publisher for any injury and/or damage to persons or property arising from any methods, products, instructions, ideas or otherwise contained in this publication.

This publication is designed to provide accurate and authoritative information with regard to the subject matter cover herein. It is sold with the clear understanding that the Publisher is not engaged in rendering legal or any other professional services. If legal, medical or any other expert assistance is required, the services of a competent person should be sought. FROM A DECLARATION OF PARTICIPANTS JOINTLY ADOPTED BY A COMMITTEE OF THE AMERICAN BAR ASSOCIATION AND A COMMITTEE OF PUBLISHERS.

Library of Congress Cataloging-in-Publication Data
Available upon request

ISBN 13 978-1-60021-211-6

Published by Nova Science Publishers, Inc. ✦New York

CONTENTS

PREFACE

In law, intellectual property (IP) is a broad entitlement extending the concept of property beyond its materialist definitions to include information, ideas, and other intangible assets in their expressed form. Depending on jurisdiction, IP rights generally enable the holder to exercise exclusive rights over the subject matter (or "work") of the IP using copyright, patent, trademark, industrial design right and similar legal protections. The term reinforces that such works are the product of the mind or intellect and that any rights may be protected at law in the same way as any other form of property. The basic public policy rationale for the protection of intellectual property is that IP laws facilitate and encourage the pursuit and disclosure of innovation into the public domain for the common good, by granting authors and inventors exclusive rights to exploit their works and invention for a limited period. However, various schools of thought are critical of the term "intellectual property." The critics characterize IP as intellectual protectionism. There is ongoing debate as to whether IP laws truly operate to confer the stated public benefits, and whether the protection they are said to provide is appropriate in the context of innovation derived from such things as traditional knowledge and folklore, and patents for software and business methods. Manifestations of this controversy can be seen in the way different jurisdictions decide whether to grant intellectual property protection. This book explores these issues as they apply to various contexts.

Chapter 1 - Congressional interest in patent policy and possible patent reform has expanded as the importance of intellectual property to innovation has increased. Patent ownership is perceived as an incentive to the technological advancement that leads to economic growth. However, growing interest in patents has been accompanied by persistent concerns about the fairness and effectiveness of the current system. Several recent studies, including those by the National Academy of Sciences and the Federal Trade Commission, have recommended patent reform to address perceived deficiencies in the operation of the patent regime. Other experts maintain that major alterations in existing law are unnecessary and that the patent process can, and is, adapting to technological progress.

The Patent Act of 2005, H.R. 2795, introduced in June 2005, would work significant legal reforms to the patent system. Among the more notable of these changes are a shift to a first-inventor-to-file priority system; substantive and procedural modifications to the patent law doctrines of willful infringement and inequitable conduct; and adoption of post-issuance opposition proceedings, prior user rights, and pre-issuance publication of all pending applications. Several of these proposals have been the subject of discussion within the patent community for many years, but others are more novel propositions.

Pending legislation attempts to address several issues of concern including the quality of issued patents, the expense and complexity of patent litigation, harmonization of U.S. patent law with the laws of our leading trading partners, potential abuses committed by patent speculators, and the special needs of individual inventors, universities, and small firms with respect to the patent system. In addition, although the existing patent statute in large measure applies the same basic rules to different sorts of inventions, regardless of the technological field of that invention, the patent system is widely believed to impact different industries in varying ways.

The provisions of H.R. 2795 would arguably work the most sweeping reforms to the U.S. patent system since the nineteenth century. However, many of these proposals, such as pre-issuance publication, prior user rights, and oppositions, have already been implemented in U.S. law to a more limited extent. These and other reforms, such as the first-inventor-to-file priority system and elimination of the best mode requirement, also reflect the decades-old patent practices of Europe, Japan, and our other leading trading partners.

Other knowledgeable observers are nonetheless concerned that certain of these proposals would weaken the patent right, thereby diminishing needed incentives for innovation. Some also believe that changes of this magnitude, occurring at the same time, do not present the most prudent course for the patent system. Patent reform therefore confronts Congress with difficult legal, practical, and policy issues, but also with apparent possibilities for altering and possibly improving the legal regime that has long been recognized as an engine of innovation within the U.S. economy.

Chapter 2 - Congress has identified research and development (R&D) as important contributors to technological progress. The performance of R&D may have intellectual property ramifications, however. To the extent that researchers use patented inventions without authorization, they may face infringement liability. Although the courts recognize an exception to patent infringement known as the "experimental use privilege," this judicially created doctrine has been described as very narrow and rarely applied. In particular, the experimental use privilege applies only to uses done for amusement, to satisfy idle curiosity or for strictly philosophical inquiry. This doctrine does not excuse uses that are in keeping with the accused infringer's business objectives.

In 2002, the U.S. Court of Appeals for the Federal Circuit applied these principles in the case of *Madey v. Duke University*. The court held that the experimental use privilege does not apply to activities that are "in keeping with the alleged infringer's legitimate business" — even though the business of the defendant, Duke University, was nonprofit research. This ruling has raised concerns among some representatives of universities and research institutions, who fear that their basic R&D activities will subject them to patent infringement lawsuits.

Competing views have arisen over the significance of the *Madey v. Duke University* case. Some commentators believe that a limited experimental use privilege may best encourage technological advancement by rewarding successful researchers with robust patent rights. Others argue that the restricted nature of the experimental use privilege may in fact limit researcher access to state-of-the-art technologies and thus discourage further technological development. Still others assert that this issue is not of great practical importance, as few patent owners will likely file costly and time-consuming lawsuits against researchers who are not making commercially important uses of patented inventions.

The judicially created, "common law" experimental use privilege is complemented by a limited statutory experimental use privilege for patents on pharmaceuticals, medical devices, and certain other products regulated by the Food and Drug Administration. This provision, enacted as part of the 1984 Hatch-Waxman Act, applies to firms seeking to market generic equivalents of brand-name products. In addition, Congress has enacted other intellectual property legislation that incorporates provisions shielding researchers from infringement liability.

Should congressional interest continue in this area, a variety of options are available. If the current scope of the common law experimental use privilege is deemed to be appropriate, then no action need be taken. Alternatively, Congress could enact legislation confirming the limited experimental use privilege recognized in *Madey v. Duke University* and predecessor cases. Introduction of a broader form of the experimental use privilege into U.S. patent law is an additional possibility.

Chapter 3 - The term "open source" refers to a computer program that is distributed along with a license, or contract, that requires users of the program to comply with specified conditions. Among these stipulations are that the source code be distributed along with the software, and that others be allowed to modify the source code as they desire. In contrast, the source code of "closed source" software is proprietary, not publicly distributed and subject to alteration only by the software manufacturer.

Some concerns have arisen concerning the relationship between open source software and intellectual property rights, including copyrights, patents and trade secrets. Although a particular computer program may be designated as open source, it remains possible that an owner of intellectual property may enforce its rights against open source software developers and users. Some commentators have also expressed concern that open source licenses may overreach, converting proprietary programs into open source software even if only a portion of that program was derived from an open source original. Others have suggested that open source licenses may not be legally enforceable, which would allow users to obtain and assert intellectual property rights pertaining to software that was initially distributed as open source.

Striking a balance between promoting innovation, on one hand, and accommodating the demands of software developers and users, on the other, forms an important component of contemporary software policy. The possibility of intellectual property rights, and their attendant license fees and royalties, may provide a significant incentive for firms to innovate and to distribute software. However, some proponents of open source software believe that these incentives are unnecessary, and further hope to maintain a non-proprietary environment of software distribution and development.

Should Congress have an interest in this area, several options present themselves. No action need be taken if the current relationship between open source software and intellectual property is deemed satisfactory, particularly as software publishers become increasingly aware of intellectual property and as judicial precedents may make the legal situation clearer. Congress might also assist individuals in identifying intellectual property that pertains to software that has been identified as open source; speak to the enforceability of open source licenses; and, as a possible more far-reaching legal reform, allow proprietary software publishers a grace period for removing portions of program code that derived from an open source original.

Chapter 4 - The United States has recently entered into a number of Free Trade Agreements, or FTAs, with Australia, Chile, Singapore, and other trading partners. Negotiations are currently ongoing with respect to the establishment of additional FTAs. In keeping with a congressional directive established in the Bipartisan Trade Promotion Act of 2002, P.L. 107-210, one objective of forming the FTAs is to establish "a standard of [intellectual property] protection similar to that found in United States law." As a result, most of the FTAs stipulate minimum levels of protection with respect to copyrights, data protection, patents, trademarks, and other forms of intellectual property. These standards relate to such provisions as the term of protection, scope of rights, and mechanisms by which these intellectual property rights are acquired and enforced.

The different FTAs vary in their comprehensiveness and level of detail. Each of these agreements has nonetheless been drafted in a manner that complies with current U.S. law. As a result, the effect of each FTA is to obligate signatories to such agreements to amend their intellectual property laws to match or resemble those of the United States.

The FTAs have been described as an effective mechanism for advancing U.S. interests in securing intellectual property protection. Increased levels of intellectual property protection with respect to computer software, music, motion pictures, and pharmaceuticals may promote a more favorable balance of trade for U.S. industry, decrease domestic prices for innovative goods and services, and serve other policy goals. The FTA framework has at times proven to be a more advantageous forum for achieving the intellectual property goals of the United States than multilateral settings.

Nonetheless, concerns have arisen over the intellectual property provisions of the FTAs. Some observers believe that certain FTA provisions may lock the United States into current intellectual property policies, inhibiting opportunities for future reform. Other commentators are concerned that under existing multilateral agreements, in particular those of the World Trade Organization, the intellectual property obligations found within one FTA may extend beyond that particular treaty partner. Finally, some observers perceive the FTAs to be an inappropriate and unfair vehicle for international intellectual property reforms due to the strong bargaining position of the United States. The scope of these potential consequences counsels continued congressional attention towards the FTAs that the United States has already formed, as well as those FTAs that are planned for the future.

Chapter 5 - Congressional interest in the development of bioterrorism countermeasures remains strong, even after passage of legislation establishing Project BioShield. In the 109th Congress, several bills have been introduced, including S. 3, the Protecting America in the War on Terror Act, and S. 975, the Project Bioshield II Act, that would generate additional incentives for the creation of new technologies to counteract potential biological threats. These bills propose reforms to current policies and practices associated with intellectual property, particularly patents, and the marketing of pharmaceuticals and related products.

Patent ownership appears to be important in the promotion of innovation, particularly in the pharmaceutical sector. Patent title provides a limited-time monopoly over the use of a discovery in exchange for the public dissemination of information contained in the patent application. This permits the inventor to receive a return on the expenditure of resources but does not guarantee that the patent will generate commercial benefits. The requirement for patent publication is expected to stimulate additional innovation to meet similar demands in the marketplace.

Currently, the Bayh-Dole Act and the Hatch-Waxman Act include provisions that utilize patent ownership to facilitate the development and commercialization of new pharmaceuticals. The Hatch-Waxman Act also contains FDA marketing approval policies that are designed to promote the creation of new drugs. Similar market-exclusivity provisions are contained in the Orphan Drug Act.

S. 3 and S. 975 would allow for the restoration of that portion of the patent term used during the FDA approval process, and/or the extension of a patent term to reward technological innovation in the area of bioterrorism countermeasures. Pending legislation would also provide for additional FDA-administered marketing exclusivities for eligible and designated countermeasures.

Encouraging the development of new counterterrorism technologies and ensuring affordable access to new drugs and medical devices are both significant goals. These aspirations may potentially conflict, however. Introducing augmented patent- and exclusivity-based incentives may stimulate innovative firms to engage in the R&D for new countermeasures, as well as to shepherd these products through time-consuming and costly marketing approval procedures. Commentators have expressed concern, however, over whether such heightened protections for innovators will be in proportion with the risks and costs of developing new countermeasures. Striking a balance between encouraging the development of new countermeasures and maintaining the traditional goals of our public health system is a central concern of the current discussion with respect to homeland security.

Chapter 6 - In *Merck KGaA v. Integra Lifesciences I, Ltd.*, __ U.S. __, 125 S. Ct. 2372 (2005), the United States Supreme Court unanimously held that the preclinical use of patented inventions in drug research is exempted from patent infringement claims by the "safe harbor" provision of the Patent Act, 35 U.S.C. § 271(e)(1). (Merck KGaA is a German company unaffiliated with the U.S.-based pharmaceutical company Merck & Co.) This decision potentially may help expedite the development of new medical treatments and lower the cost of some drugs for consumers.

In 2003, the U.S. Court of Appeals for the Federal Circuit had narrowly construed the safe harbor provision as protecting only clinical research activities that produce information for submission to the Food and Drug Administration (FDA) in the regulatory process. In vacating that decision, the U.S. Supreme Court ruled that the exemption applies to *all* uses of patented inventions that are "reasonably related" to the process of developing any information for FDA submission. The Court explained that, under certain conditions, the safe harbor provision is even "sufficiently broad" to protect the use of patented compounds in experiments that are not ultimately submitted to the FDA or drug experiments that are not ultimately the subject of an FDA submission. Finally, the scope of the exemption is not limited only to preclinical studies pertaining to a drug's safety in humans, but also includes preclinical data regarding a drug's efficacy, mechanism of action, pharmacokinetics, and pharmacology.

However, the Court cautioned that the exemption does not reach all experimental activity that at some point, however attenuated, may lead to an FDA approval process. For example, the safe harbor provision does not embrace basic scientific research performed on a patented compound without the intent to develop a particular drug or without a reasonable belief that the compound will cause a particular physiological effect that the researcher desires. In addition, because the matter was not at issue in the case, the Court expressly declined to decide whether or to what extent the exemption applies to patented "research tools" that are often used to facilitate general research in developing compounds for FDA submissions.

Chapter 7 - The Digital Millennium Copyright Act (DMCA) prohibits individuals from manufacturing, selling, or trafficking in technology, products, services, or devices that circumvent technology designed to control access to a copyrighted work. This is known as the DMCA's "anticircumvention" provision. Although most commonly invoked in the context of digital piracy of music, motion pictures, and other entertainment-related media, another genre of anticircumvention-based cases is making its way through the courts. These cases involve the initiation of anticircumvention litigation for what some argue are anti-competitive purposes in the marketing and sale of durable goods.

The practice of reverse engineering allows others to identify and analyze the creative versus the functional aspects of copyrighted software and to utilize them to some degree. In some contexts, reverse engineering has been held by the courts to be a fair use of copyright-protected property. There is also an express, limited statutory exemption for reverse engineering under the DMCA. How the practice and use of reverse engineering for commercial goals relates to the relatively new protections against circumvention is of interest to many. They are concerned with the extent to which access-control technology is likely be employed to extend (or attempt to extend) a copyright holder's control over durable goods with copyrighted components and the secondary markets for such goods.

This report examines two recent decisions from U.S. Courts of Appeals, *Lexmark International, Inc. v. Static Control Components, Inc.*, 387 F.3d 522 (6th Cir. 2004), and *Chamberlain Group, Inc. v. Skylink Technologies, Inc.*, 381 F.3d 1178 (Fed. Cir. 2004). Both cases allege violation of the anticircumvention statute with respect to development and sale of consumer goods. In *Lexmark*, the defendant marketed a microchip that allowed third-party manufacturers to sell toner cartridges that worked with the plaintiff's printer. In *Chamberlain*, the defendant sold a universal garage door opener transmitter that worked with the plaintiff's garage door opener. In both cases, the courts found that the actions of the defendants did *not* violate the anticircumvention provisions of the DMCA. In doing so, the courts had to reconcile closely related aspects of copyright law with the strictures of the DMCA.

Chapter 8 - Digital Rights Management (DRM) refers to the technology that copyright owners use to protect digital media. This report surveys several of the DRM bills that were introduced in the 107th Congress and those that are pending in the 108th Congress. Generally, the bills are directed at two separate goals. One goal is to increase access to digitally-protected media for lawful purposes. The other attempts to thwart digital piracy and would do so by enhancing civil and criminal sanctions for digital (and traditional) copyright infringement and educating the public about the rights of copyright holders.

Although no bills were enacted during the 107th Congress, two of the bills focusing on access have been reintroduced in the 108th Congress. Representatives Boucher and Lofgren reintroduced their bills from the 107th Congress. They are H.R. 107, the "Digital Media Consumers' Rights Act of 2003" and H.R. 1066, respectively. H.R. 1066 is renamed the "Benefit Authors without Limiting Advancement or Net Consumer Expectations (BALANCE) Act of 2003." And Senator Wyden introduced S. 692, a labeling disclosure bill entitled the "Digital Consumers Right to Know Act."

S. 1621, the "Consumers, Schools, and Libraries Digital Management Awareness Act of 2003," addresses several DRM issues. It would prohibit the Federal Communication Commission from establishing mandatory technology standards and require disclosure requirements for access controlled digital media and consumer electronics. It also addresses the subpoena process by which copyright owners acquire personal information about suspected infringers.

Bills addressing piracy include H.R. 4077, the "Piracy Deterrence and Education Act of 2004"; S. 1932, the "Artists Rights and Theft Prevention Act of 2004"; and, S. 2237, the "Protecting Intellectual Rights Against Theft and Appropriation Act of 2004." S. 1932 and S. 2237 have been reported favorably by the Senate Judiciary Committee. Because digital transmission poses the greatest distribution risk to entertainment content owners, these bills attempt to thwart the initial unauthorized copying and/or uploading to the Internet. Hence, sanctions for illegal distribution of pre-release commercial works and surreptitious recording of movies in theaters are emphasized.

Chapter 9 - This report addresses the adoption of a "broadcast flag" system by the Federal Communications Commission (FCC) to protect digital television (DTV) broadcasts from unauthorized redistribution. The report also addresses the recent decision of the United States Court of Appeals for the District of Columbia Circuit reversing and vacating the FCC's broadcast flag report and order.

Chapter 10 - Google, Inc. is digitally scanning the collections of several prominent libraries in order to create a vast searchable database of literary works. Copyright holders who have not authorized and object to the digitization have filed suit against the company. This report provides background on the pending litigation. It will be updated as judicial developments warrant.

Complaints for copyright infringement were recently filed against Google, Inc. by a variety of authors and representatives of the book publishing industry.[1] The complaints specifically challenge Google's "Print Library" project, recently renamed "Google Book Search," an effort by Google in conjunction with several library partners to scan books into a digital format so that they may be searched textually. Although the case is in its very early stages, the issues presented have captured national attention. Once again, new technology and traditional principles of copyright law appear to be in conflict. Because of the unique facts and issues presented, there is scant legal precedent to legitimize Google's claim that its project is protected by copyright law's fair use exception to liability for infringement.[2] Thus, questions presented may be ones of first impression for the courts.

In: New Intellectual Property Issues
Editor: John V. Luong, pp. 1-36
ISBN: 978-1-60021-211-6
© 2007 Nova Science Publishers, Inc.

Chapter 1

PATENT REFORM: INNOVATION ISSUES*

Wendy H. Schacht and John R. Thomas

SUMMARY

Congressional interest in patent policy and possible patent reform has expanded as the importance of intellectual property to innovation has increased. Patent ownership is perceived as an incentive to the technological advancement that leads to economic growth. However, growing interest in patents has been accompanied by persistent concerns about the fairness and effectiveness of the current system. Several recent studies, including those by the National Academy of Sciences and the Federal Trade Commission, have recommended patent reform to address perceived deficiencies in the operation of the patent regime. Other experts maintain that major alterations in existing law are unnecessary and that the patent process can, and is, adapting to technological progress.

The Patent Act of 2005, H.R. 2795, introduced in June 2005, would work significant legal reforms to the patent system. Among the more notable of these changes are a shift to a first-inventor-to-file priority system; substantive and procedural modifications to the patent law doctrines of willful infringement and inequitable conduct; and adoption of post-issuance opposition proceedings, prior user rights, and pre-issuance publication of all pending applications. Several of these proposals have been the subject of discussion within the patent community for many years, but others are more novel propositions.

Pending legislation attempts to address several issues of concern including the quality of issued patents, the expense and complexity of patent litigation, harmonization of U.S. patent law with the laws of our leading trading partners, potential abuses committed by patent speculators, and the special needs of individual inventors, universities, and small firms with respect to the patent system. In addition, although the existing patent statute in large measure applies the same basic rules to different sorts of inventions, regardless of the technological field of that invention, the patent system is widely believed to impact different industries in varying ways.

The provisions of H.R. 2795 would arguably work the most sweeping reforms to the U.S. patent system since the nineteenth century. However, many of these proposals, such as pre-issuance publication, prior user rights, and oppositions, have already been implemented in U.S. law to a more limited extent. These and other reforms, such as the first-inventor-to-file priority system and elimination of the best mode requirement, also reflect the decades-old patent practices of Europe, Japan, and our other leading trading partners.

* Excerpted from CRS Report RL32996, Updated July 15, 2005

Other knowledgeable observers are nonetheless concerned that certain of these proposals would weaken the patent right, thereby diminishing needed incentives for innovation. Some also believe that changes of this magnitude, occurring at the same time, do not present the most prudent course for the patent system. Patent reform therefore confronts Congress with difficult legal, practical, and policy issues, but also with apparent possibilities for altering and possibly improving the legal regime that has long been recognized as an engine of innovation within the U.S. economy.

INTRODUCTION

Congressional interest in patent reform has increased as the patent system becomes more significant to U.S. industry. There is broad agreement that more patents are sought and enforced then ever before; that the attention paid to patents in business transactions and corporate boardrooms has dramatically increased; and that the commercial and social significance of patent grants, licenses, judgments, and settlements is at an all-time high.[1] As the United States becomes even more of a high-technology, knowledge-based economy, the importance of patents may grow even further in the future.

Increasing interest in patents has been accompanied by persistent concerns about the fairness and effectiveness of the current system. Several recent studies, including those by the National Academy of Sciences and the Federal Trade Commission, have recommended patent reform to address perceived deficiencies in the operation of the patent regime.[2] Other experts maintain that major alterations in existing law are unnecessary and that the patent process can, and is, adapting to technological progress.

In June 2005, legislation was introduced that attempts to respond to current concerns about the functioning of the patent process. The Patent Act of 2005, H.R. 2795, proposes significant legal reforms to the patent system, including a shift to a first-inventor-to-file priority system; substantive and procedural modifications to the patent law doctrines of willful infringement and inequitable conduct; and adoption of post-issuance opposition proceedings, prior user rights, and pre-issuance publication of all pending applications. Several of these proposals have been the subject of discussion within the patent community for many years, but others are more novel propositions.

This report provides an overview of current patent reform issues. It begins by offering a summary of the structure of the current patent system and the role of patents in innovation policy. The report then reviews some of the broader issues and concerns, including patent quality, the high costs of patent litigation, international harmonization, and speculation in patents, that have motivated H.R. 2795's diverse reform proposals. The specific components of this legislation are then identified and reviewed in greater detail.

PATENTS AND INNOVATION POLICY

The Mechanics of the Patent System

The patent system is grounded in Article I, Section 8, Clause 8 of the U.S. Constitution, which states that "The Congress Shall Have Power . . . To promote the Progress of Science and useful Arts, by securing for limited Times to Authors and Inventors the exclusive Right to their respective Writings and Discoveries . . . ." As mandated by the Patent Act of 1952,[3] U.S. patent rights do not arise automatically. Inventors must prepare and submit applications to the U.S. Patent and Trademark Office (USPTO) if they wish to obtain patent protection.[4] USPTO officials known as examiners then assess whether the application merits the award of a patent.[5] The patent acquisition process is commonly known as "prosecution."[6]

In deciding whether to approve a patent application, a USPTO examiner will consider whether the submitted application fully discloses and distinctly claims the invention.[7] In addition, the application must disclose the "best mode," or preferred way, that the applicant knows to practice the invention.[8] The examiner will also determine whether the invention itself fulfills certain substantive standards set by the patent statute. To be patentable, an invention must be useful, novel and nonobvious. The requirement of usefulness, or utility, is satisfied if the invention is operable and provides a tangible benefit.[9] To be judged novel, the invention must not be fully anticipated by a prior patent, publication or other knowledge within the public domain.[10] A nonobvious invention must not have been readily within the ordinary skills of a competent artisan at the time the invention was made.[11]

If the USPTO allows the patent to issue, the patent proprietor obtains the right to exclude others from making, using, selling, offering to sell or importing into the United States the patented invention.[12] Those who engage in these acts without the permission of the patentee during the term of the patent can be held liable for infringement. Adjudicated infringers may be enjoined from further infringing acts.[13] The patent statute also provides for the award of damages "adequate to compensate for the infringement, but in no event less than a reasonable royalty for the use made of the invention by the infringer."[14]

The maximum term of patent protection is ordinarily set at 20 years from the date the application is filed.[15] At the end of that period, others may employ that invention without regard to the expired patent.

Patent rights are not self-enforcing. Patentees who wish to compel others to observe their rights must commence enforcement proceedings, which most commonly consist of litigation in the federal courts. Although issued patents enjoy a presumption of validity, accused infringers may assert that a patent is invalid or unenforceable on a number of grounds.[16] The U.S. Court of Appeals for the Federal Circuit (Federal Circuit) possesses national jurisdiction over most patent appeals from the district courts.[17] The U.S. Supreme Court enjoys discretionary authority to review cases decided by the Court of Appeals for the Federal Circuit.[18]

Innovation Policy

Patent ownership is perceived to be an incentive to innovation, the basis for the technological advancement that contributes to economic growth. It is through the commercialization and use of new products and processes that productivity gains are made and the scope and quality of goods and services are expanded. Award of a patent is intended to stimulate the investment necessary to develop an idea and bring it to the marketplace embodied in a product or process. Patent title provides the recipient with a limited-time monopoly over the use of his discovery in exchange for the public dissemination of information contained in the patent application. This is intended to permit the inventor to receive a return on the expenditure of resources leading to the discovery but does not guarantee that the patent will generate commercial benefits. The requirement for publication of the patent is expected to stimulate additional innovation and other creative means to meet similar and expanded demands in the marketplace.

Innovation produces new knowledge. One characteristic of this knowledge is that it is a "public good," a good that is not consumed when it is used. This "public good" concept underlies the U.S. patent system. Absent a patent system, "free riders" could easily duplicate and exploit the inventions of others. Further, because they incurred no cost to develop and perfect the technology involved, copyists could undersell the original inventor. The resulting inability of inventors to capitalize on their inventions would lead to an environment where too few inventions are made.[19] The patent system corrects this market failure problem by providing innovators with an exclusive interest in their inventions, thereby allowing them to capture its marketplace value.

The regime of patents purportedly serves other goals as well. The patent system encourages the disclosure of products and processes, for each issued patent must include a description sufficient to enable skilled artisans to practice the patented invention.[20] At the close of the patent's twenty-year term,[21] others may practice the claimed invention without regard to the expired patent. In this manner the patent system ultimately contributes to the growth of the public domain.

Even during their term, issued patents may also encourage others to "invent around" the patentee's proprietary interest. A patentee may point the way to new products, markets, economies of production and even entire industries. Others can build upon the disclosure of a patent instrument to produce their own technologies that fall outside the exclusive rights associated with the patent.[22]

The patent system has also been identified as a facilitator of markets. Absent patent rights, an inventor may have scant tangible assets to sell or license. In addition, an inventor might otherwise be unable to police the conduct of a contracting party. Any technology or know-how that has been disclosed to a prospective licensee might be appropriated without compensation to the inventor. The availability of patent protection decreases the ability of contracting parties to engage in opportunistic behavior. By lowering such transaction costs, the patent system may make technology-based transactions more feasible.[23]

Through these mechanisms, the patent system can act in more socially desirable ways than its chief legal alternative, trade secret protection. Trade secrecy guards against the improper appropriation of valuable, commercially useful and secret information. In contrast to patenting, trade secret protection does not result in the disclosure of publicly valuable information. That is because an enterprise must take reasonable measures to keep secret the information for which trade secret protection is sought. Taking the steps necessary to maintain secrecy, such as implementing physical security measures, also imposes costs that may ultimately be unproductive for society.[24]

The patent system has long been subject to criticism, however. Some observers have asserted that the patent system is unnecessary due to market forces that already suffice to create an optimal level of innovation. The desire to obtain a lead time advantage over competitors, as well as the recognition that technologically backward firms lose out to their rivals, may well provide sufficient inducement to invent without the need for further incentives.[25] Other commentators believe that the patent system encourages industry concentration and presents a barrier to entry in some markets.[26] Still other observers believe that the patent system too frequently attracts speculators who prefer to acquire and enforce patents rather than engage in socially productive activity.[27]

When analyzing the validity of these competing views, it is important to note the lack of rigorous analytical methods available for studying the effect of the patent law upon the U.S. economy as a whole. The relationship between innovation and patent rights remains poorly understood. As a result, current economic and policy tools do not allow us to calibrate the patent system precisely in order to produce an optimal level of investment in innovation. Thus, each of the arguments for and against the patent system remains open to challenge by those who are unpersuaded by their internal logic.

CURRENT ISSUES AND CONCERNS

Pending legislation proposes a number of changes to diverse aspects of the patent system. Although these reforms are undoubtedly motivated by a range of concerns, a discrete number of issues have been the subject of persistent discussion in the patent community over a period of many years. Among these issues are concern for the quality of issued patents, the expense and complexity of patent litigation, harmonization of U.S. patent law with the laws of our leading trading partners, potential abuses committed by patent speculators, and the special needs of individual inventors, universities, and small firms with respect to the patent system. In addition, although the patent statute in large measure applies the same basic rules to different sorts of inventions, regardless of the technological field of that invention, the patent system is widely believed to impact different industries in varying ways.[28] As a result, different industries can be expected to espouse different views of certain patent reform proposals. Before turning to a more specific analysis of individual legislative proposals, this report reviews the proposed legislation's broader themes with regard to these issues and concerns.

Patent Quality

Government, industry, academia and the patent bar alike have long insisted that the USPTO approve only those patent applications that describe and claim a patentable advance.[29] Because they meet all the requirements imposed by the patent laws, quality patents may be dependably enforced in court and employed as a technology transfer tool. Such patents are said to confirm private rights by making their proprietary uses, and therefore their value, more predictable. Quality patents also may clarify the extent that others may approach the protected invention without infringing. These traits in turn should strengthen the incentives of private actors to engage in value-maximizing activities such as innovation or commercial transactions.[30]

In contrast, poor patent quality is said to hold deleterious consequences. Large numbers of inappropriately granted patents may negatively impact entrepreneurs. For example, innovative firms may be approached by an individual with a low quality patent that appears to cover the product they are marketing. The innovative firm may recognize that the cost of challenging a patent even of dubious validity may be considerable. Therefore, the firm may choose to make payments under licensing arrangements, or perhaps decide not to market its product at all, rather than contest the patent proprietor's claims.[31]

Poor patent quality may also encourage opportunistic behavior. Perhaps attracted by large damages awards and a potentially porous USPTO, rent-seeking entrepreneurs may be attracted to form speculative patent acquisition and enforcement ventures. Industry participants may also be forced to expend considerable sums on patent acquisition and enforcement.[32] The net results would be reduced rates of innovation, decreased patent-based transactions, and higher prices for goods and services.

Although low patent quality appears to affect both investors and competitors of a patentee, patent proprietors themselves may also be negatively impacted. Patent owners may make managerial decisions, such as whether to build production facilities or sell a product, based upon their expectation of exclusive rights in a particular invention. If their patent is declared invalid by the USPTO or a court, the patentee will be stripped of exclusive rights without compensation. The issuance of large numbers of invalid patents would increase the possibility that the investment-backed expectations of patentees would be disappointed.[33]

The notion that high patent quality is socially desirable has been challenged, however. Some commentators believe that market forces will efficiently assign patent rights no matter what their quality. Others observe that few issued patents are the subject of litigation and further estimate that only a minority of patents are licensed or sold. Because many patented inventions are not used in a way that calls their validity into question, some observers maintain, society may be better off making a detailed review into the patentability of an invention only in those few cases where that invention is of commercial significance.[34]

Proposed legislation bears upon the patent quality issue. In particular, H.R. 2795 would allow for increased public participation in USPTO decisionmaking through a pre-issuance submission procedure. This bill would also allow for post-issuance opposition proceedings, which would potentially allow interested parties to "weed out" invalid patents before they are the subject of licensing or infringement litigation.

Litigation Costs

Patent enforcement is often expensive. The complex legal and technological issues, extensive discovery proceedings, expert witnesses, and specially qualified attorneys associated with patent trials can lead to high costs.[35] One study published in 2000 concluded that the average cost of patent enforcement was $1.2 million.[36] These expenses appear to be increasing, with one more recent commentator describing an "industry rule of thumb" whereby "any patent infringement lawsuit will easily cost $1.5 million in legal fees alone to defend."[37] Higher stakes litigation is even more costly: For patent suits involving damages claims of more than $25 million, expenses reportedly increase to $4 million per side.[38]

For innovative firms that are not infrequently charged with patent infringement, or that bring claims of patent infringement themselves, the annual expenses associated with patent litigation can be very dear. The Microsoft Corporation reportedly defends an average of 35 to 40 patent lawsuits annually at a cost of almost $100 million.[39] The Intel Corporation has recently been estimated to spend $20 million a year on patent litigation.[40]

The high costs of litigation may discourage patent proprietors from bringing meritorious claims against infringers. They may also encourage firms to license patents of dubious merit rather than contest them in court. Pending legislation endeavors to make patent litigation less costly and complex through modification of the patent law doctrines of willful infringement and inequitable conduct. It would also call for an administrative opposition proceeding that, in some measure, could serve as a less expensive alternative to litigation.

International Harmonization

In our increasingly globalized, high-technology economy, patent protection in a single jurisdiction is often ineffective to protect the interests of inventors. As a result, U.S. inventors commonly seek patent protection abroad. Doing so can be a costly, time-consuming, and difficult process. There is no global patent system. Inventors who desire intellectual property protection in a particular country must therefore take specific steps to procure a patent within that jurisdiction.[41]

Differences in national laws are among the difficulties faced by U.S. inventors seeking patent rights overseas. Although the world's patent laws have undergone considerable harmonization in recent years, several notable distinctions between U.S. patent law and those of our leading trading partners persist. Pending legislation would address some of these differences by modifying U.S. patent law in order to comply with international standards. Among these proposed reforms are adoption of a first-inventor-to-file priority system, a post-issuance opposition system, assignee filing, publication of all pending patent applications, and prior user rights; elimination of the best mode requirement; and encouragement for the adoption of a one-year grace period within the European Patent Convention and Japanese Patent Act.

Potential Abuses of Patent Speculators

Some commentators believe that the patent system too frequently attracts speculators who prefer to acquire and enforce patents rather than engage in research, development, manufacturing, or other socially productive activity.[42] Patent speculators are sometimes termed "trolls," after creatures from folklore that would emerge from under a bridge in order to waylay travelers.[43] The late Jerome C. Lemelson, a prolific inventor who owned hundreds of patents and launched numerous charges of patent infringement, has sometimes been mentioned in this context. The annual revenue of the Lemelson estate's patent licensing program has been reported as in excess of $1.5 billion.[44] But as explained by journalist Michael Ravnitsky, "critics charge that many Lemelson patents are so-called submarine patents, overly broad applications that took so long to issue or were so general in nature that their owners could unfairly claim broad infringement across entire industry sectors."[45] Of such patent ventures, patent attorney James Pooley observes:

> Of course there is nothing inherently wrong with charging someone rent to use your property, including intellectual property like patents. But it's useful to keep in mind — especially when listening to prattle about losing American jobs to foreign competition — that these patent mills produce no products. Their only output is paper, of a highly threatening sort.[46]

Patent enforcement suits brought by patent speculators appear to present special concerns for manufacturers and service providers. If one manufacturer or service provider commences litigation against another, the defendant can often counter with its own claims of patent infringement against the plaintiff. Because patent speculators do not otherwise participate in the marketplace, however, they are immune to such counterclaims. This asymmetry in litigation positions reportedly reduces the bargaining power of manufacturers and service providers and exposes them to harassment.[47]

Observers hasten to note, however, that not every patent proprietor who does not commercialize the patented invention should properly be considered an opportunistic "troll." A nonmanufacturing patentee may lack the expertise or resources to produce a patented product, prefer to commit itself to further innovation, or otherwise have legitimate reasons for its behavior.[48] Universities and small biotechnology companies often fit into this category. Further, whether classified as a "troll" or not, each patent owner has presumptively fulfilled all of the relevant statutory requirements. Among these obligations is a thorough disclosure of a novel, nonobviousn invention to the public.[49]

Pending legislation would impact concerns over "trolling" by amendment to the patent statute's provision regarding injunctions, as well as allowing limitations upon so-called continuation practice at the USPTO.

The Role of Individuals, Universities and Small Entities

Entrepreneurs and small, innovative firms play a role in the technological advancement and economic growth of the United States.[50] Several studies commissioned by U.S. federal agencies have concluded that individuals and small entities constitute a significant source of innovative products and services.[51] Studies have also indicated that entrepreneurs and small, innovative firms rely more heavily upon the patent system than larger enterprises. Larger companies are said to possess alternative means for achieving a proprietary or property-like interest in a particular technology. For example, trade secrecy, ready access to markets, trademark rights, speed of development, and consumer goodwill may to some degree act as substitutes to the patent system.[52] However, individual inventors and small firms often do not have these mechanisms at their disposal. As a result, the patent system may enjoy heightened importance with respect to these enterprises.[53]

In recent years, universities have also become more full-fledged participants in the patent system. This trend has been attributed to the Bayh-Dole Act,[54] a federal statute that allowed universities and other government contractors to retain patent title to inventions developed with the benefit of federal funding.[55] In recent years there has reportedly "been a dramatic increase in academic institutions' investments in technology licensing activities."[56] This increase has been reflected in the growth in the number of patents held by universities, the number of universities with technology transfer offices, and the amount of patent-based licensing revenues that these offices have raised.[57]

The U.S. patent system have long acknowledged the role, and particular needs, of independent inventors, small firms, and universities. For example, the patent statute calls for each of these entities to receive a 50% discount on many USPTO fees.[58] As the USPTO is currently entirely funded by the fees it charges its users,[59] this provision effectively calls for larger institutions to subsidize the patent expenditures of their smaller competitors.

Beyond potentially diminished financial resources vis-a-vis larger concerns, however, observers have disagreed over whether independent inventors, small firms, and universities have particular needs with respect to the patent system, and if so whether those needs should be reflected in patent law doctrines. With respect to the proposed system of "prior user rights,"[60] for example, some observers state that such rights would particularly benefit small entities, which may often lack a sophisticated knowledge of the patent system.[61] Others disagree, stating that smaller concerns rely heavily on the exclusivity of the patent right, and that the adoption of prior user rights would advantage large enterprises.[62] Similar debates have occurred with respect to other patent reform proposals, perhaps reflecting the fact that the community of independent inventors, small firms, and universities is itself a diverse one.

Provisions of pending legislation that appear to be of particular interest to independent inventors, universities, and small businesses include a shift to a first-inventor-to-file priority system, prior user rights, pre-issuance publication of all pending patent applications, and post-issuance oppositions.

Different Roles for Patents in Distinct Industries

To a large extent, the patent statute subjects all inventions to the same standards, regardless of the field in which those inventions arose. Whether the invention is an automobile engine, semiconductor, or a pharmaceutical, it is for the most part subject to the same patentability requirements, scope of rights, and term of protection. Both experience and economic research suggest that distinct industries experience the patent system in different ways, however.[63] As a result, it can be expected that particular industries will react differently to the various patent reform proposals currently before Congress.

Although broad generalizations should be drawn with care, two industries widely perceived as viewing the patent system in different ways are the pharmaceutical and software sectors. Within the pharmaceutical industry, individual patents are perceived as critical to a business model that provides life-saving and life-enhancing medical innovations, but eventually allows members of the public access to medicines at low cost. In particular, often only a handful, and sometimes only one or two patents cover a particular drug product. Patents are also judged to be crucial to the pharmaceutical sector because of the relative ease of replicating the finished product. For example, while it is expensive, complicated, and time consuming to duplicate an airplane, it is relatively simple to perform a chemical analysis of a pill and reproduce it.[64]

In contrast to the pharmaceutical field, the nature of software development is such that innovations are typically cumulative and new products often embody numerous patentable inventions. This environment has led to what has been described as a poor match between patents and products in the [software] industry: it is difficult to patent an entire product in the software industry because any particular product is likely to include dozens if not hundreds of separate technological ideas.[65]

This situation may be augmented by the multiplicity of patents often associated with a finished computer product that utilizes the software. It is not uncommon for thousands of different patents (relating to hardware and software) to be embodied in one single computer.

In addition, ownership of these patents may well be fractured among hundreds or thousands of different individuals and firms.

In summary, then, the patent laws provide a "one size fits all" system, where all inventions are subject to the same requirements of patentability and scope of protection, regardless of the technical field in which they arose. Innovators in different fields nonetheless have varying experiences with the patent system. These discrepancies, among others, lead to the expectation that distinct industries may react differently to the various patent reform proposals presently considered by Congress.

PROPOSED LEGISLATIVE INITIATIVES

Pending legislation proposes a diverse array of patent reforms. The remainder of this report identifies and reviews these proposals.

First Inventor to File

Pending legislation would alter the U.S. patent priority rule from the current "first-to-invent" principle to the "first-inventor-to-file" principle.[66] Within the patent law, the priority rule addresses the circumstance where two or more persons independently develop the identical or similar invention at approximately the same time. In such cases the patent law must establish a rule as to which of these inventors obtains entitlement to a patent.[67]

In the United States, when more than one patent application is filed claiming the same invention, the patent will be awarded to the applicant who was the first inventor in fact. This conclusion holds even if the first inventor was not the first person to file a patent application directed towards that invention.[68] Under this "first-to-invent" system,[69] the timing of real-world events, such as the date a chemist conceived of a new compound or a machinist constructed a new engine, is of significance.

In every patent-issuing nation except the United States, priority of invention is established by the earliest effective filing date of a patent application disclosing the claiming invention.[70] Stated differently, the inventor who first filed an application at the patent office is presumptively entitled to the patent. Whether or not the first applicant was actually the first individual to complete the invention in the field is irrelevant. This priority system follows the "first-inventor-to file" principle.

A simple example illustrates the distinction between these priority rules. Suppose that inventor A synthesizes a new chemical compound on August 1, 2005, and files a patent application on November 1, 2005 claiming that compound. Suppose further that inventor B independently invents the same compound on September 1, 2005, and files a patent application on October 1, 2005. Inventor A would be awarded the patent under the first-to-invent rule, while Inventor B would obtain the patent under the first-inventor-to-file principle.

Under the current U.S. first-to-invent rule, the majority of priority disputes in the United States are resolved via "interference" proceedings conducted at the USPTO.[71] An interference is a complex administrative proceeding that ordinarily results in the award of priority to one of its participants. These proceedings are not especially common. One estimate is that less than one-quarter of one percent of patents are subject to an interference.[72] This statistic may mislead, however, because the expense of interference cases may result in their use only for the most commercially significant inventions.

The patent community has witnessed an extensive and sometimes emotional debate on the relative merits of the first-to-invent and first-inventor-to-file principle. Supporters of the current first-to-invent principle in part assert that the first-inventor-to-file system would create inequities by sponsoring a "race to the Patent Office."

They are also concerned that the first-to-file system would encourage premature and sketchy technological disclosures in hastily-filed patent applications.[73]

Supporters of the first-inventor-to-file principle in part assert that it provides a definite, readily determined and fixed date of priority of invention, which would lead to greater legal certainty within innovative industries. They also contend that the first-inventor-to-file principle would decrease the complexity, length and expense associated with current USPTO interference proceedings. Rather than being caught up in lengthy interference proceedings in an attempt to prove dates of inventive activity that occurred many years previously, they assert, inventors could continue to go about the process of innovation. Supporters also observe that informed U.S. firms already organize their affairs on a first-inventor-to-file basis in order to avoid forfeiture of patent rights abroad.[74]

The effect of a shift to the first-inventor-to-file rule upon individual inventors, small firms, and universities has been debated. Some observers state that such entities often possess fewer resources and wherewithal than their larger competitors, and thus are less able to prepare and file patent applications quickly. Others disagree, stating that smaller concerns are more nimble than larger ones and thus better able to submit applications promptly. They also point to the availability of provisional applications,[75] asserting that such applications allow small entities to secure priority rights readily without a significant expenditure of resources. A quantitative study of interference proceedings by Gerald Mossinghoff, a former Commissioner of the USPTO, also suggested that the first-to-invent rule neither advantaged nor disadvantaged small entities vis-a-vis larger enterprises.[76]

The role of the U.S. Constitution is sometimes debated within the context of the patent priority principle. Article I, section 8, clause 8 of the Constitution provides Congress with the authority to award "inventors" with exclusive rights. Some observers suggest this language suggests, or possibly even mandates, the current first-to-invent system. Others conclude that because the first-inventor-to-file only awards patents to individuals who actually developed the invention themselves, rather than derived it from another, this priority system is permissible under the Constitution.[77]

In weighing the validity of this position, it should be noted that under well-established U.S. law, the first-inventor-in-fact does not always obtain entitlement to a patent. If, for example, a first-inventor-in-fact maintained his invention as a trade secret for many years before seeking patent protection, he may be judged to have "abandoned, suppressed or concealed" the invention.[78] In such a case a second-inventor-in-fact may be awarded a patent on that invention. Courts have reasoned that this statutory rule encourages individuals to disclose their inventions to the public promptly, or give way to an inventor who in fact does so.[79] As the first-inventor-to-file rule acts in a similar fashion to this longstanding patent law principle, conflict between this rule and the Constitution appears unlikely.

Notably, a first-inventor-to-file priority rule does not permit one individual to copy another's invention and then, by virtue of being the first to file a patent application, be entitled to a patent. All patent applicants must have originated the invention themselves, rather than derived it from another.[80] In order to police this requirement, H.R. 2795 provides for "inventor's rights contests" that would allow the USPTO to determine which applicant is entitled to a patent on a particular invention.[81]

Grace Period

Pending legislation would also impact the existing one-year "grace period" enjoyed by U.S. inventors. Current U.S. patent law essentially provides inventors with a one-year period to decide whether patent protection is desirable, and, if so, to prepare an application. Specified activities that occur before the "critical date" —patent parlance for the day one year before the application was filed — will prevent a patent from issuing.[82] If, for example, an entrepreneur first discloses an invention by publishing an article in a scientific journal, she knows that she has one year from the publication date in which to file a patent application. Importantly, uses, sales, and other technical disclosures by third parties will also start the one-year clock running. As a result, inventors have a broader range of concerns than merely their own behavior.[83]

In contrast, many other patent-granting states provide more limited grace periods, or no grace periods at all. In Europe, any sales or publication of an invention anywhere in the world prior to the filing date defeats the patentability of an invention.[84] The Japanese patent system includes a six-month grace period tied only to the activities of the inventor.[85] Under the patent law of Japan, any disclosures of an invention made by a third party even one day before the filing date will prevent the issuance of a patent.

The proposed legislation would make two changes to U.S. patent law. First, the one-year grace period would apply only to the inventor's own activities. Third party activity that occurred even one day before the filing date would constitute prior art and potentially be patent-defeating. This rule represents a change from current law, under which a patent applicant may possibly "antedate" prior art by showing that she invented the subject matter of the application prior to the date of the third party reference.[86] This change is consistent with the proposed shift to a first-inventor-to-file priority rule from a first-to-invent system.

Second, the proposed legislation would include a provision that might encourage adoption of a one-year grace period for inventor activities in Europe and Japan. An understanding of this proposal requires some background information on the international priority system established by the Paris Convention.[87] The international priority system allows an inventor to file a patent application in one Paris Convention signatory state, which is usually the inventor's home country. If the inventor subsequently files patent applications in any other Paris Convention signatory state within the next 12 months, overseas patent-granting authorities will treat the application as if it were filed on the first filing date. Critically, information that enters the public domain between the priority date and subsequent filing dates does not prejudice the later applications. Paris Convention priority allows U.S. inventors to preserve their original USPTO filing dates as they make arrangements to file patent applications overseas.[88]

Suppose, for example, that an inventor files a patent application at the USPTO on October 1, 2005. The inventor then files a patent application claiming the same invention in the Japanese Patent Office on September 1, 2006. As part of his Japanese application, the inventor informs the Japanese Patent Office of the earlier U.S. application. Because Japan has acceded to the Paris Convention, the Japanese Patent Office will treat that inventor's application as if it had been filed on October 1, 2005. As a result, information that entered the public domain after the U.S. filing date would not prejudice the inventor's Japanese application. A journal article published on January 1, 2006, for example, would not limit the opportunity of the inventor to obtain a Japanese patent.

The U.S. patent statute currently limits the usefulness of the Paris Convention priority date for foreign inventors seeking U.S. patent rights. Section 119 of the Patent Act states that:

> no patent shall be granted on any application for patent for an invention which had been patented or described in a printed publication in any country more than one year before the date of the actual filing of the application in this country, or which had been in public use or on sale in this country more than one year prior to such filing.[89]

The effect of this language is that the one-year grace period is measured not from the Paris Convention international priority date, but the actual U.S. filing date.

This limitation may discourage U.S. trading partners from adopting a grace period analogous to that of U.S. law. Consider, for example, a Japanese inventor who publishes an article in a scientific journal describing his new invention on August 1, 2004. Consistent with Japanese patent law, he then files a patent application at the Japanese Patent Office six months later, on February 1, 2005. Then, in accordance with the Paris Convention, he files an application at the USPTO on February 1, 2006.

Under these circumstances, the U.S. patent application should be denied, even though the Japanese inventor appeared to comply with all legal formalities. Because the U.S. patent statute compels the USPTO to assess the grace period as ending as the actual U.S. filing date in 2006, rather than the Paris Convention priority date in 2005, the U.S. patent is barred from issuance. This state of affairs may give pause to nations considering adopting a U.S.-style grace period. Foreign applicants who rely upon grace periods within their own national systems may be put in a position of forfeiture of their U.S. patent rights.

Apparently aware of this concern, pending legislation potentially changes the date the grace period closes from the actual U.S. filing date to the Paris Convention priority date — provided that Europe and Japan adopt laws analogous to that of proposed U.S. law. In the language of H.R. 2795:

> Before the date, if ever, that the Director of the [USPTO] publishes a notice . . . declaring that both the European Patent Convention and the patent laws of Japan afford inventors seeking patents a 1-year period prior to the effective filing date of a claimed invention during which disclosures made by the inventor or by others who obtained the subject matter disclosed directly or indirectly from the inventor do not constitute prior art, the term "effective filing date" as used in section 102(a)(1)(A) of title 35, United States Code, shall be construed by disregarding any right of priority except that provided under section 119(e) of title 35, United States Code.[90]

Should the USPTO Director publish a notice in keeping with this provision, then foreign inventors would be able to rely upon their domestic grace periods and maintain their ability to obtain patents in the United States.

Elimination of Sections 102(c), (d) and (f)

Pending legislation would eliminate three provisions of the Patent Act, paragraphs (c), (d), and (f) of Section 102. Section 102(c) does not allow an applicant to obtain a patent when he "has abandoned the invention." This statute does not refer to disposal of the invention itself, however, but instead to the intentional surrender of an invention *to the public*. Older Supreme Court opinions instruct that abandonment may occur where an inventor expressly dedicates it to the public, through a deliberate relinquishment or conduct evidencing an intent not to pursue patent protection.[91] The circumstances must be such that others could reasonably rely upon the inventor's renunciation.[92] Perhaps because few individuals expressly cede their patentable inventions to the public without seeking compensation, there are few modern judicial opinions that consider 35 U.S.C. § 102(c) in any meaningful way. In addition, the generally applicable principle of equitable estoppel may apparently be used to obtain the same result.[93]

Like section 102(c), section 102(d) of the Patent Act is reportedly little-used.[94] 35 U.S.C. 102(d) bars a U.S. patent when (1) an inventor files a foreign patent application more than twelve months before filing the U.S. application, and (2) a foreign patent results from that application prior to the U.S. filing date. Suppose that an inventor files an application at a foreign patent office on May 25, 2004. The foreign application matures into a granted foreign patent on August 1, 2005. If the inventor has not filed his patent application at the USPTO as of August 1, 2005, the date of the foreign patent grant, any patent application that the inventor subsequently filed in the United States would be defeated.

The policy basis for 35 U.S.C. § 102(d) is to encourage the prompt filing of patent applications in the United States. As the Patent Office Commissioner explained in 1870:

> The intention of [C]ongress obviously was to obtain for this country the free use of the inventions of foreigners as soon as they became free abroad. This is indicated by the use of the phrase, 'first patented, or caused to be patented, in a foreign country,' for it was presumable that American citizens would obtain their first patent here, while a foreigner would first patent his invention in his own country. The statute was designed to prevent a foreigner from spending his time and capital in the development of an invention in his own country, and then coming to this country to enjoy a further monopoly, when the invention had become free at home. The result of such a course would be that while the foreign country was developing the invention and enjoying its benefits, its use could be interdicted here; while, if the term of the monopoly could be further extended here, the market could be controlled long after the foreign nation was prepared to flood this country with the unpatented products of the patented process.[95]

Section 102(d) has been subject to critical commentary. Because inventors may choose to file a patent application only in the United States, the policy goal of assuring that the U.S. market will become patent-free contemporaneously with foreign markets may not be well-served by this provision. In addition, 35 U.S.C. § 102(d) effectively acts against foreign, rather than U.S.-based inventors, as domestic inventors ordinarily file at the USPTO first before seeking rights overseas. Some commentators have suggested that 35 U.S.C. § 102(d) violates the spirit, if not the letter, of U.S. international treaty obligations, which generally impose an obligation of national treatment with respect to intellectual property matters.[96]

Finally, H.R. 2795 would also eliminate current 35 U.S.C. § 102(f), which states that a person may obtain a patent unless "he did not himself invent the subject matter sought to be patented." As this requirement that only an actual inventor may obtain a patent is also stated by 35 U.S.C. § 101,[97] this proposed amendment does not appear to have any substantive impact upon the patent law.

Assignee Filing

Under current U.S. law, a patent application must be filed by the inventor —that is to say, the natural person or persons who developed the invention.[98] This rule applies even where the invention was developed by individuals in their capacity as employees. Even though rights to the invention have usually been contractually assigned to an employer, for example, the actual inventor, rather than the employer, must be the one that applies for the patent. Section 118 of the Patent Act allows a few exceptions to this general rule. If an inventor cannot be located, or refuses to perform his contractual obligation to assign an invention to his employer, then the employer may file in place of the inventor.[99]

H.R. 2795 would instead stipulate that a "person to whom the inventor has assigned or is under an obligation to assign the invention may make an application for patent."[100] Individuals who otherwise make a showing of a "sufficient proprietary interest in the matter" may also apply for a patent on behalf of the inventor upon a sufficient show of proof of the pertinent facts. Under the proposed legislation, if the USPTO "Director grants a patent on an application filed under this section by a person other than the inventor, the patent shall be granted to the real party in interest and upon such notice to the inventor as the Director considers to be sufficient."

Legal reforms allowing assignee filing of patent applications have been discussed for many years. A 1966 Report of the President's Commission on the Patent System recommended this change as a way to simplify formalities of application filing and of avoiding the delays caused by the need to identify and obtain signatures from each inventor.[101] The 1992 Advisory Commission on Patent Law Reform was also in favor of this change. The 1992 Commission observed that the United States was "the only country which does not permit the assignee of an invention to file a patent application in its own name."[102] In the opinion of the 1992 Commission, assignee filing would appropriately accompany a U.S. shift to a first-inventor-to-file priority system, as the reduction of formalities would allow innovative enterprises to more promptly file patent applications.

The 1992 Commission also reviewed potential undesirable aspects of assignee filing. The Commission noted that patent applications filed by assignees may lack the actual inventor's personal guarantee that the application was properly filed. In addition, assignee filing might derogate the right of natural persons to their inventions. In the opinion of the Commission, however, the advantages of assignee filing outweighed the disadvantages.[103]

Elimination of the Best Mode Requirement

H.R. 2795 would eliminate U.S. patent law's best mode requirement. Currently, inventors are required to "set forth the best mode contemplated by the inventor of carrying out his invention." [104] Failure to disclose the best mode known to the inventor is a ground for invalidating an issued patent. The courts have established a two-part standard for analyzing whether an inventor disclosed her best mode in a particular patent. The first inquiry was whether the inventor knew of a way of practicing the claimed invention that he considered superior to any other. If so, then the patent instrument must identify, and disclose sufficient information to enable persons of skill in the art to practice that best mode.[105]

Proponents of the best mode requirement have asserted that it allows the public to receive the most advantageous implementation of the technology known to the inventor. This disclosure becomes part of the patent literature and may be freely reviewed by those who wish to design around the patented invention. Members of the public are also said to be better able to compete with the patentee on equal footing after the patent expires.[106]

The best mode requirement has encountered severe criticism in recent years, however.[107] For example, a 1992 Presidential Commission recommended that Congress eliminate the best mode requirement. The Commission reasoned that patents are also statutorily required to disclose "the manner and process of making and using [the invention], in such full, clear, concise, and exact terms as to enable any person skilled in the art . . . to make and use the same."[108] This "enablement" requirement was believed to provide sufficient information to achieve the patent law's policy goals.[109]

The Commission further stated that the best mode requirement leads to increases in the costs and complexity of patent litigation. As the Commission explained:

> The disturbing rise in the number of best mode challenges over the past 20 years may serve as an indicator that the best mode defense is being used primarily as a procedural tactic. A party currently can assert failure to satisfy the best mode requirement without any significant burden. This assertion also entitles the party to seek discovery on the "subjective beliefs" of the inventors prior to the filing date of the patent application. This broad authority provides ample opportunity for discovery abuse. Given the fluidity by which the requirement is evaluated (e.g., even accidental failure to disclose any superior element, setting, or step can negate the validity of the patent), and the wide ranging opportunities for discovery, it is almost certain that a best mode challenge will survive at least initial judicial scrutiny.[110]

The Commission further reasoned that the best mode at the time of filing is unlikely to remain the best mode when the patent expires many years later.[111] Because many foreign patent laws include no analog to the best mode requirement, inventors based overseas have also questioned the desirability of the best mode requirement in U.S. law.

Inequitable Conduct

The administrative process of obtaining a patent from the USPTO has traditionally been conducted as an *ex parte* procedure. Stated differently, patent prosecution involves only the applicant and the USPTO. Members of the public, and in particular the patent applicant's marketplace competitors, do not participate in patent acquisition procedures.[112] As a result, the patent system relies to a great extent upon applicant observance of a duty of candor and truthfulness towards the USPTO.

An applicant's obligation to proceed in good faith may be undermined, however, by the great incentive applicants possess not to disclose, or to misrepresent, information that might deleteriously impact their prospective patent rights. The patent law therefore penalizes those who stray from honest and forthright dealings with the USPTO. Under the doctrine of "inequitable conduct," if an applicant intentionally misrepresents a material fact or fails to disclose material information, then the resulting patent will be declared unenforceable.[113] Two elements must exist before a court will decide that the applicant has engaged in inequitable conduct. First, the patentee must have misrepresented or failed to disclose material information to the USPTO in the prosecution of the patent.[114] Second, such nondisclosure or misrepresentation must have been intentional.[115]

During patent infringement litigation, an accused infringer has the option of asserting that the plaintiff's patent is unenforceable because it was procured through inequitable conduct. Concerns have arisen that charges of inequitable conduct have become routine in patent cases. As one commentator explains:

> The strategic and technical advantages that the inequitable conduct defense offers the accused infringer make it almost too attractive to ignore. In addition to the potential effect on the outcome of the litigation, injecting the inequitable conduct issue into patent litigation wreaks havoc in the patentee's camp. The inequitable conduct defense places the patentee on the defensive, subjects the motives and conduct of the patentee's personnel to intense scrutiny, and provides an avenue for discovery of attorney-client and work product documents...[116]

As the Federal Circuit put it, "the habit of charging inequitable conduct in almost every major patent case has become an absolute plague."[117] Other observers believe that because inequitable conduct requires an analysis of the knowledge and intentions of the patent applicants, the doctrine may also be contributing disproportionately to the time and expense of patent litigation.[118]

Due to these perceived burdens upon patent litigation, some commentators have proposed that the inequitable conduct defense be eliminated.[119] Others believe that inequitable conduct is necessary to ensure the proper functioning of the patent system. As the Advisory Commission on Patent Law Reform explained in its 1992 report:

> Some mechanism to ensure fair dealing between the patentee, public, and the Federal Government has been part of the patent system for over 200 years. In its modern form, the unenforceability defense provides a necessary incentive for patent applicants to engage in fair and open dealing with the [USPTO] during the ex parte prosecution of patent applications, by imposing the penalty of forfeiture of patent rights for failure to so deal. The defense is also considered to be an essential safeguard against truly fraudulent conduct before the [USPTO]. Finally, the defense provides a means for encouraging complete disclosure of information relevant to a particular patent application. . . . . Thus, from a policy perspective, the defense of unenforceability based upon inequitable conduct is desirable and should be retained.[120]

Proposed legislation would retain the concept of an inequitable conduct defense, but introduce a number of substantive and procedural changes to the doctrine. In broad overview, H.R. 2795 would provide statutory authorization for the USPTO Director to issue regulations governing applicants' duty of candor.[121] Although the USPTO has for many years issued such regulations,[122] commentators have questioned the authority of the USPTO to do so under current law.[123]

H.R. 2795 would also provide the USPTO with authority to prosecute violations of the inequitable conduct doctrine. Although that agency previously performed this function without express statutory authorization, the USPTO suspended that activity in 1992.[124] Under H.R. 2795, the USPTO would be obligated to establish a "special office with authority to investigate possible violations of the duty of candor and good faith." The USPTO would be granted subpoena power and be able to conduct hearings in pursuit of these investigations. The USPTO would also be able to penalize misconduct through substantial civil fines.

In addition, H.R. 2795 would limit the circumstances under which the defense of inequitable conduct could be raised before the courts. In broad outline, under H.R. 2795 if a court determines an issue of possible misconduct, then the court is directed to refer the matter to the USPTO. Within judicial infringement proceedings, issues of inequitable conduct could only arise after the court grants a motion to amend the pleadings. Such a motion would have to describe the relevant facts in detail and could not be granted until the court has previously entered a judgment that at least one of the asserted patent claims is invalid.

H.R. 2795 would also introduce a substantive change to the law of inequitable conduct. Under this legislation, a finding of inequitable conduct is appropriate only in circumstances when the USPTO examiner relied upon the alleged misconduct. In particular, the charge of inequitable conduct cannot be sustained unless the USPTO "would not have issued the invalidated claim, acting reasonably, in the absence of the misconduct," or "based upon the prosecution history as a whole objectively considered, would have done so based upon in whole or in part on account of the misconduct."[125]

Publication of Pending Applications

Until recent years, the U.S. patent system maintained pending patent applications in secrecy. The first moment that the public would become aware of the existence of a U.S. patent application was the day the USPTO formally allowed it to issue as a granted patent. This regime advantaged patent applicants because it allowed them to understand exactly what the scope of any allowed claims might be prior to disclosing an invention. Thus, if the applicant was able to maintain the invention that was subject to a patent application as a trade secret, then he could choose between obtaining the allowed patent claims and trade secret status. In addition, because the invention was not disclosed prior to the award of formal patent rights, unscrupulous competitors were discouraged from copying the invention.

However, this secrecy regime has been perceived as imposing costs as well. Others might well engage in duplicative research efforts during the pendency of patent applications, unaware that an earlier inventor had already staked a claim to that technology. This arrangement also allowed inventors to commence infringement litigation on the very day a patent issued, without any degree of notice to other members of the technological community.[126]

Industry in the United States possessed one mechanism for identifying pending U.S. patent applications. Most foreign patent regimes publish all pending patent applications approximately 18 months after they have been filed.[127] As a result, savvy firms in the United States could review pending applications filed before foreign patent offices, and make an educated guess as to the existence of a corresponding U.S. application. This effort was necessarily inexact, however, particularly as some inventors either lacked the resources, or made the strategic decision, not to obtain patent rights outside the United States.

In enacting the American Inventors Protection Act of 1999, Congress for the first time introduced the concept of pre-grant publication into U.S. law. Since November 29, 2000, U.S. patent applications have been published 18 months from the date of filing, with some exceptions. The most significant of these exceptions applies where the inventor represents that he will not seek patent protection abroad. In particular, if an applicant certifies that the invention disclosed in the U.S. application will not be the subject of a patent application in another country that requires publication of applications 18 months after filing, then the USPTO will not publish the application.[128] As a result, inventors who do not wish to seek foreign patent rights retain the possibility of avoiding pre-grant publication.

Proposed legislation would further modify the U.S. pre-grant publication system by effectively calling for all pending applications to be published approximately 18 months after they are filed. In particular, H.R. 2795 would eliminate the possibility of opting out of pre-grant publication by certifying that a patent will be sought only in the United States.[129]

Pre-issuance Submissions

Pending legislation would expand the ability of members of the public to submit information to the USPTO that is pertinent to pending applications. Under current law, interested individuals may enter a protest against a patent application. The protest must specifically identify the application and be served upon the applicant. The protest must also include a copy and, if necessary, an English translation, of any patent, publication or other information relied upon. The protester also must explain the relevance of each item.[130]

Protest proceedings have traditionally played a small role in U.S. patent practice. Until Congress enacted the American Inventors Protection Act of 1999, the USPTO maintained applications in secrecy. Therefore, the circumstances in which members of the public would learn of the precise contents of a pending patent application were relatively limited. With the USPTO commencing publication of some pending patent applications, protests would seem far more likely. Seemingly aware of this possibility, the 1999 Act provided that the USPTO shall "ensure that no protest or other form of pre-issuance opposition . . . may be initiated after publication of the application without the express written consent of the applicant."[131] Of course, the effect of this provision is to eliminate the possibility of protest in exactly that class of cases where the public is most likely to learn of the contents of a pending application.

Through rulemaking, the USPTO has nonetheless established a limited mechanism for members of the public to submit information they believe is pertinent to a pending, published application. The submitted information must consist of either a patent or printed publication, and it must be submitted within two months of the date the USPTO published the pending application. Nondocumentary information that may be relevant to the patentability determination, such as sales or public use of the invention, will not be considered.[132] In addition, because Congress stipulated that no protest or pre-grant opposition may occur absent the consent of the patent holder, the USPTO has explained that it will not accept *comments* or *explanations* concerning the submitted patents or printed publications. If such comments are attached, USPTO staff will redact them before the submitted documents are forwarded to the examiner.[133]

Pending legislation would augment the possibility for pre-issuance submissions. Under H.R. 2795, any person would be able to submit patent documents and other printed publications to the USPTO for review. Such prior art must be submitted within the later date of either (1) the date the USPTO issues a notice of allowance to the patent applicant; or (2) either six months after the date of pre-grant publication of the application, or the date of the rejection of any claim by the USPTO examiner. Such a submission must include "a concise description of the asserted relevance of each submitted document."[134]

Most observers agree that ideally, the USPTO would have access to all pertinent information when making patentability determinations. A more expansive pre-issuance submission policy may allow members of the public to disclose relevant patents and other documents that the USPTO's own searchers may not have revealed, thereby leading to more accurate USPTO decision making. On the other hand, lengthy pre-issuance submissions may merely be repetitive of the USPTO's own search results, but still require extensive periods of examiner review that might ultimately delay examination. The proposal of H.R. 2795 apparently attempts to balance these concerns by expanding existing opportunities for post-publication submissions, but limiting the timing and nature of those submissions so as to prevent undue burdens upon the USPTO and patent applicants.

Continuation Applications

Pending legislation would allow the USPTO Director to limit the availability of so-called continuation applications via regulation. Continuation applications essentially allow inventors to re-file previously rejected patent applications in order to pursue further prosecution with a USPTO examiner.[135] The filing of a patent application effectively allows two formal communications, termed "Office Actions," with a USPTO examiner. Agreement often cannot be reached by this point, however, leaving the applicant with only the alternatives of abandonment of patent protection or the filing of an appeal. By filing a continuation application, an applicant essentially purchases additional time for dialogue between the applicant and examiner.

The use of continuation applications is commonplace in U.S. patent practice. Applicants not infrequently file one or more continuing applications based upon an earlier filed "patent" application. Many patents have issued based upon chains of continuation applications involving a parent, grandparent, and even more remote predecessors.

Continuation applications are said to allow applicants to more accurately claim a previously disclosed invention without the necessity of an appeal.[136] Some commentators believe they are subject to abuse, however. Under this view, continuation practice introduces delay and uncertainty into the patent acquisition process. In particular, applicants are said to use a chain of continuation applications in order to gain advantages over competitors by waiting to see what product the competitor will make, and then drafting patent claims that cover that product. Continuation practice is also said to have led to long delays in the issuance of a patent in order to surprise an established industry, a process known as "submarine patenting."[137]

At present time, two features of the Patent Act have been identified as mitigating against abuse of continuation applications. The first feature, discussed above, is the pre-grant publication of pending patent applications.[138] This rule allows interested members of the public to be notified that the application has been filed at the USPTO prior to the formal bestowal of exclusive patent rights. Under current law, however, exceptions exist to this rule, and as a result some applications are not published prior to their issuance.[139]

The second feature is the term of the patent, which as of June 8, 1995, was set to the duration of twenty years from the date of filing.[140] The twenty-year term is measured from the filing date of the earliest in the chain of continuation applications. Inventors who file a series of continuation applications thereby shorten their term of eventual patent protection. In some circumstances, however, even a patent with a diminished term may remain of competitive significance. And again there is an exception to the twenty-year patent term based upon the filing date: Applications filed prior to June 8, 1995, may instead enjoy a term of seventeen years set from the date of issuance.

H.R. 2795 would allow the USPTO Director to issue regulations that would limit the ability of inventors to file continuation applications. The bill would further stipulate that "[n]o such regulation may delay applicants an adequate opportunity to obtain claims for any invention disclosed in an application for patent."[141]

Prior User Rights

Pending legislation would expand the applicability of a "first inventor defense" established by the American Inventors Protection Act of 1999. As currently found at 35 U.S.C. § 273, an earlier inventor of a "method of doing or conducting business" that was later patented by another may claim a defense to patent infringement in certain circumstances. H.R. 2795 would broaden this defense by allowing it to apply with respect to any patented subject matter.

The impetus for this provision lies in the rather complex relationship between the law of trade secrets and the patent system. Trade secrecy protects individuals from misappropriation of valuable information that is useful in commerce. One reason an inventor might maintain the invention as a trade secret rather than seek patent protection is that the subject matter of the invention may not be regarded as patentable. Such inventions as customer lists or data compilations have traditionally been regarded as amenable to trade secret protection but not to patenting.[142] Inventors might also maintain trade secret protection due to ignorance of the patent system or because they believe they can keep their invention as a secret longer than the period of exclusivity granted through the patent system.[143]

The patent law does not favor trade secret holders, however. Well-established patent law provides that an inventor who makes a secret, commercial use of an invention for more than one year prior to filing a patent application at the USPTO forfeits his own right to a patent.[144] This policy is based principally upon the desire to maintain the integrity of the statutorily prescribed patent term. The patent law grants patents a term of twenty years, commencing from the date a patent application is filed.[145] If the trade secret holder could make commercial use of an invention for many years before choosing to file a patent application, he could disrupt this regime by delaying the expiration date of his patent.

On the other hand, settled patent law principles established that prior secret uses would not defeat the patents of later inventors.[146] If an earlier inventor made secret commercial use of an invention, and another person independently invented the same technology later and obtained patent protection, then the trade secret holder could face liability for patent infringement. This policy is based upon the reasoning that once issued, published patent instruments fully inform the public about the invention, while trade secrets do not. As between a subsequent inventor who patented the invention, and thus had disclosed the invention to the public, and an earlier trade secret holder who had not, the law favored the patent holder.

An example may clarify this rather complex legal situation. Suppose that Inventor A develops and makes commercial use of a new manufacturing process. Inventor A chooses not to obtain patent protection, but rather maintains that process as a trade secret. Many years later, Inventor B independently develops the same manufacturing process and promptly files a patent application claiming that invention. In such circumstances, Inventor A's earlier, trade secret use does not prevent Inventor B from procuring a patent. Furthermore, if the USPTO approves the patent application, then Inventor A faces infringement liability should Inventor B file suit against him.

The American Inventors Protection Act of 1999 somewhat modified this principle. That statute in part provided an infringement defense for an earlier inventor of a "method of doing or conducting business" that was later patented by another. By limiting this defense to patented methods of doing business, Congress responded to the 1998 Federal Circuit opinion in *State Street Bank and Trust Co. v. Signature Financial Group*.[147] That judicial opinion recognized that business methods could be subject to patenting, potentially exposing individuals who had maintained business methods as trade secrets to liability for patent infringement.

Again, an example may aid understanding of the first inventor defense. Suppose that Inventor X develops and exploits commercially a new method of doing business. Inventor X maintains his business method as a trade secret. Many years later, Inventor Y independently develops the same business method and promptly files a patent application claiming that invention. Even following the enactment of the American Inventors Protection Act, Inventor X's earlier, trade secret use would not prevent Inventor Y from procuring a patent. However, should the USPTO approve Inventor Y's patent application, and should Inventor Y sue Inventor X for patent infringement, then Inventor X may potentially claim the benefit of the first inventor defense. If successful,[148] Inventor X would enjoy a complete defense to infringement of Inventor Y's patent.

As originally enacted, the first inventor defense applied only to patents claiming a "method of doing or conducting business." Although the American Inventors Protection Act did not define this term, the first inventor defense was arguably a focused provision directed towards a specific group of potential patent infringers.

Pending legislation would expand upon the first inventor defense by allowing it to apply to all patented subject matter.[149] By removing current restrictions referring to methods of doing business, H.R. 2795 would effectively introduce "prior user rights" into U.S. law.

A feature of many foreign patent regimes, prior user rights are often seen as assisting small entities, which may lack the sophistication or resources to pursue patent protection. The provision of prior user rights would allow such entities to commercialize their inventions when they used the subject matter of the invention prior to the patent's filing date, even when they themselves did not pursue patent rights. For this reason, a more expansive prior user rights regime has also been tied to adoption of the first-inventor-to-file priority system.[150]

Proponents of prior user rights also assert that the new legislation would support investment in technological innovation. Under this view, firms would not longer be required to engage in extensive defensive patenting, but rather would be able to devote these resources to further innovation. In addition, some commentators observe that many U.S. trading partners, including Germany and Japan, currently allow prior user rights. As a result, U.S. firms that obtain patent rights in certain foreign nations may face the possibility that a foreign firm may enjoy prior user rights in that invention. Foreign firms with U.S. patents do not currently face this possibility with respect to U.S. firms, however. Under this view, adoption of prior user rights in the United States would "level the playing field" for U.S. industry.[151]

Proposals to adopt prior user rights have attracted critics, however. Some observers believe that this regime would benefit large corporations at the expense of smaller ones. Others believe that individuals who are aware that they can rely upon prior user rights will be less likely to disclose their inventions through the patent system. Still others have stated that prior user rights reduce the value of patents and therefore make innovation less desirable. The role of the U.S. Constitution is sometimes debated within this context as well. Article I, section 8, clause 8 of the Constitution provides Congress with the authority to award "inventors the exclusive right to their . . . discoveries." Some commentators suggest this language suggests, or possibly requires, a system of exclusive patent rights, rather than an interest that may be mitigated by prior user rights.[152]

Injunctions

35 U.S.C. § 283 currently allows courts to "grant injunctions in accordance with the principles of equity to prevent the violation of any right secured by patent, on such terms as the court deems reasonable."[153] H.R. 2795 would add the following additional language to that statute:

> In determining equity, the court shall consider the fairness of the remedy in light of all the facts and the relevant interests of the parties associated with the invention. Unless the injunction is entered pursuant to a nonappealable judgment of infringement, a court shall stay the injunction pending an appeal upon an affirmative showing that the stay would not result in irreparable harm to the owner of the patent and that the balance of the hardships from the stay does not favor the owner of the patent.[154]

In other words, this legislation would stipulate that all the facts and relevant interests must be considered when a court weighs "principles of equity" in deciding whether to enjoin an adjudicated infringer or not.

Understanding of this proposed reform requires a review of current judicial precedent concerning the availability of injunctions in patent cases. Under current law, courts ordinarily grant permanent injunctions to patentees that prevail in infringement litigation.[155] As the Federal Circuit recently explained:

> Because the "right to exclude recognized in a patent is but the essence of the concept of property," the general rule is that a permanent injunction will issue once infringement and validity have been adjudged. To be sure, "courts have in rare instances exercised their discretion to deny injunctive relief in order to protect the public interest." *Rite-Hite Corp. v. Kelley, Inc.*, 56 F.3d 1538, 1547 (Fed. Cir.1995); *see Roche Prods., Inc. v. Bolar Pharm. Co.*, 733 F.2d 858, 865-66 (Fed. Cir.1984) ("standards of the public interest, not the requirements of private litigation, measure the propriety and need for injunctive relief"). Thus, we have stated that a court may decline to enter an injunction when "a patentee's failure to practice the patented invention frustrates an important public need for the invention," such as the need to use an invention to protect public health.[156]

As this language suggests, few published judicial opinions decline to grant injunctions against adjudicated patent infringers. The usually cited exception is the1934 decision in *City of Milwaukee v. Activated Sludge*,[157] where the Court of Appeals for the Seventh Circuit refused to enjoin infringement of a patented method for sewage treatment. Had the city of Milwaukee been prevented from using the patented invention, it would have been required to dump large quantities of raw sewage into Lake Michigan. Observing that "the health and the lives of more than half a million people are involved,"[158] the court denied the requested injunction. Because the patentee still obtained judicially determined monetary remedies against the city of Milwaukee, this outcome essentially amounted to an award of a compulsory license.

Another notable case, the 1944 decision in *Vitamin Technologists, Inc. v. Wisconsin Alumni Research Foundation*,[159] involved a patent claiming a method of irradiating foods to increase Vitamin D content. This treatment helped eliminate the debilitating Vitamin D deficiency disease called rickets. The availability of Vitamin D-enhanced margarine was particularly important to the poor, who were better able to afford margarine as compared to butter. The patent proprietor had nonetheless refused to license the patent to margarine producers. The Court of Appeals for the Ninth Circuit ultimately held the asserted patents invalid or unenforceable on conventional grounds. However, the court also discussed the concept that injunctions should be refused where they act against public health interests.[160]

Other courts have taken a different route by delaying the effective date of a permanent injunction rather than refusing it altogether. For example, in *Schneider (Europe) AG v. SciMed Life Systems, Inc.*,[161] the adjudicated infringer marketed a rapid-exchange catheter used by surgeons. Although the court concluded that no evidence of record supported a finding that the infringing product was more safe or objectively superior to other catheters on the market, the court recognized that some physicians did strongly prefer the infringing product.[162] The court opted to grant a permanent injunction with a delay of one year from the entry of judgment. The court reasoned that this year-long transition period would allow surgeons to switch from the infringing product with a minimum of disruption, at least in comparison with the immediate imposition of an injunction. The court further provided that the patentee would receive an escalating royalty during the transition period.[163]

Some commentators have expressed concerns over the current state of the law. It has been observed that although the patent statute directs courts to "grant injunctions in accordance with the principles of equity,"[164] in practice courts issue injunctions in favor of victorious patent proprietors "virtually automatically."[165] In the view of some observers, this rule has encouraged strategic behavior by speculators. These speculators do not themselves market goods or services, and thus do not acquire patents in order to protect their own markets. Rather, they are said to use patents to threaten manufacturers and service providers. Because such speculators may legitimately threaten to halt use of the patented invention entirely, accused infringers may enter into a license on even a dubious patent in the face of losing the company business. This practice is sometimes termed "trolling," after creatures from folklore that would emerge from under a bridge in order to waylay travelers.[166]

In response, some commentators have observed that not all patentees that chose not to market their patented inventions are fairly characterized as opportunistic speculators. Particular individuals or firms may lack the expertise or resources to produce a patented product, or otherwise have sound business reasons not to do so. In addition, as the Federal Circuit explained a decade ago:

> A patent is granted in exchange for a patentee's disclosure of an invention, not for the patentee's use of the invention. There is no requirement in this country that a patentee make, use, or sell its patented invention.[167]

Still other observers view patents as time-limited property rights.[168] Under this view, infringers are properly enjoined so that the patent owner's exclusive rights may be preserved. Failing to enjoin infringements may also diminish the incentives needed for innovators to develop patentable inventions in the first instance.

On the other hand, scholarly commentary concludes that the term "intellectual property" is one of relatively recent vintage, and that patents can just as readily be analogized as arising under the tort law, or as constituting a government-granted subsidy.[169] Although viewing a patent as a property right arguably suggests that an injunction is a suitable remedy for a violation of that right, these alternative conceptions of the patent grant do not so strongly imply that courts should enjoin adjudicated infringers as a matter of course. It should also be noted that even traditional properties, such as real estate, are prospectively subject to numerous limitations, including easements, zoning restrictions, servitudes, and other obligations.[170]

Limitation upon Damages

Pending legislation would also address the award of damages where the patented invention forms but one component of the infringer's larger commercial product or process. In the wording of H.R. 2795:

> In determining a reasonable royalty in the case of a combination, the court shall consider, if relevant and among other factors, the portion of the realizable profit that should be credited to the inventive contribution as distinguished from other features of the combination, the manufacturing process, business risks, or significant features or improvements added by the infringer.[171]

This proposed reform appears to be directed towards perceived concerns about overly generous damages awards in this context. As one commentator asserted:

> [I]nventors have learned to abuse the patent system and increase leverage against manufacturers by pursuing "system claims" in the [USPTO]. These clever claims insert the crux of the predator's "innovation" into much larger contexts than that to which the inventor is rightfully entitled. For example, the abuser may actually have invented a hinge mechanism, but draws the patent claim to a door including the hinge mechanism. In this example, the door is well known to, and long in use by, the public but in subsequent litigation, the patent predator claims entitlement to, and the court awards, damages based on the entire value of the door rather than on the value of the innovative hinge.[172]

Several observations may be made about this proposed reform. First, the Federal Circuit has explained that:

> Virtually *all* patents are "combination patents," if by that label one intends to describe patents having claims to inventions formed of a combination of elements. It is difficult to visualize, at

least in the mechanical-structural arts, a "non-combination" invention, i.e., an invention consisting of a *single* element. Such inventions, if they exist, are rare indeed.[173]

Under this view, this legislative proposal is of potentially broad application. Further, the required assessment of the "inventive contribution" of a patented combination —rather than base its damages determination upon the claims themselves — would mark a notable change in U.S. patent law. As the Federal Circuit recently stated:

It is well settled that "there is no legally recognizable or protected 'essential' element, gist or 'heart' of the invention in a combination patent." Rather, "'[t]he invention' is defined by the claims."[174]

Finally the language of H.R. 2795 expressly applies only to an award of a reasonable royalty. However, the patent statute also authorizes the courts to award damages equal to the lost profits of the patentee,[175] provided the patentee can make the necessary showing. It is unclear why this proposal is limited to damages awards based upon reasonable royalties, rather than upon both forms of damages awards available under the patent laws.

Willful Infringement

Pending legislation would also reform the law of willful infringement. The patent statute currently provides that the court "may increase the damages up to three times the amount found or assessed."[176] An award of enhanced damages, as well as the amount by which the damages will be increased, is committed to the discretion of the trial court.[177] Although the statute does not specify the circumstances in which enhanced damages are appropriate, the courts most commonly award them when the infringer acted in blatant disregard of the patentee's rights. This circumstance is termed "willful infringement."[178]

Courts will not ordinarily enhance damages due to willful infringement if the adjudicated infringer did not know of the patent until charged with infringement in court, or if the infringer acted with the reasonable belief that the patent was not infringed or that it was invalid. Federal Circuit decisions emphasize the duty of someone with actual notice of a competitor's patent to exercise due care in determining if his acts will infringe that patent. A common way to fulfill this obligation is to obtain competent legal advice before commencing, or continuing, activity that may infringe another's patent.[179]

Prior to 2004, the Federal Circuit held that when an accused infringer invoked the attorney-client or work-product privilege, courts should be free to reach an adverse inference that either (1) no opinion had been obtained or (2) an opinion had been obtained and was contrary to the infringers's desire to continue practicing the patented invention.[180] However, in its decision in *Knorr-Bremse Systeme fuer Nutzfahrzeuge GmbH v. Dana Corp.*,[181] the Federal Circuit expressly overturned this principle. The Court of Appeals further stressed that the failure to obtain legal advice did not occasion an adverse inference with respect to willful infringement either. Following the *Knorr-Bremse* opinion, willful infringement determinations are based upon "the totality of circumstances, but without the evidentiary contribution or presumptive weight of an adverse inference that any opinion of counsel was or would have been unfavorable."[182]

Patent law's willful infringement doctrine has proven controversial. Some observers believe that this doctrine ensures that patent rights will be respected in the marketplace. Critics of the policy believe that the possibility of trebled damages discourages individuals from reviewing issued patents. Out of fear that their inquisitiveness will result in multiple damages, innovators may simply avoid looking at patents until they are sued for infringement. To the extent this observation is correct, the law of willful infringement discourages the dissemination of technical knowledge, thereby thwarting one of the principal goals of the patent system. Fear of increased liability for willful infringement may also discourage firms from challenging patents of dubious validity. Consequently some have argued that the patent system should shift to a "no-fault" regime of strictly compensatory damages, without regard to the state of mind of the adjudicated infringer.[183]

H.R. 2795 would add several clarifications and changes to the law of willful infringement. First, a finding of willful infringement would be appropriate only where (1) the infringer received specific written notice from the patentee and continued to infringe after a reasonable opportunity to investigate; (2) the infringer intentionally copied from the patentee with knowledge of the patent; and (3) the infringer continued to infringe after an adverse court ruling. Second, willful infringement cannot be found where the infringer possessed an informed, good faith belief that its conduct was not infringing. Finally, a court may not determine willful infringement before the date on which a determination is made that the patent is not invalid, enforceable, and infringed.[184]

Post-issuance Opposition Proceedings

Pending legislation would introduce post-issuance opposition proceedings into U.S. patent law. Oppositions, which are common in foreign patent regimes, are patent revocation proceedings that are usually administered by authorities from the national patent office. Oppositions often involve a wide range of potential invalidity arguments and are conducted through adversarial hearings that resemble courtroom litigation.

Although the U.S. patent system does not currently include oppositions, the U.S. patent system has incorporated a so-called reexamination proceeding since 1981. Some commentators have viewed the reexamination as a more limited form of an opposition. Under the reexamination statute, any individual, including the patentee, a competitor, and even the USPTO Director, may cite a prior art patent or printed publication to the USPTO. If the USPTO determines that this reference raises a "substantial new question of patentability" with respect to an issued patent, then it will essentially reopen prosecution of the issued patent.

Traditional reexamination proceedings are conducted in an accelerated fashion on an *ex parte* basis. Following the American Inventors Protection Act of 1999, an *inter partes* reexamination allows the requester to participate more fully in the proceedings through the submission of arguments and the filing of appeals. Either sort of reexamination may result in a certificate confirming the patentability of the original claims, an amended patent with narrower claims or a declaration of patent invalidity.

Congress intended reexamination proceedings to serve as an inexpensive alternative to judicial determinations of patent validity.[185] Reexamination also allows further access to the legal and technical expertise of the USPTO after a patent has issued.[186] However, some commentators believe that reexamination proceedings have been employed only sparingly and question their effectiveness.[187]

Some analysts have expressed concern that potential requesters are discouraged from commencing *inter partes* reexamination proceedings due to a statutory provision that limits their future options. In order to discourage abuse of these proceedings, the *inter partes* reexamination statute provides that third-party participants may not later assert that a patent is invalid "on any ground that [they] raised or could have raised during the inter partes reexamination proceedings."[188] Some observers believe that this potential estoppel effect disinclines potential requesters from use of this post-issuance proceeding. In apparent response to this concern, H.R. 2795 would delete the phrase "or could have raised" from the statute.[189] As a result, *inter partes* reexamination requesters would be limited only with respect to arguments that they actually made before the USPTO.

In addition to reforming the reexamination statute, H.R. 2795 would create an additional post-issuance proceeding termed an "opposition." Under this bill, any person could commence an opposition either within nine months after the issuance of the patent, or six months after receiving notice from the patent holder alleging infringement. The opposition may relate to a wide range of patentability issues, including double patenting, statutory subject matter, novelty, nonobviousness, enablement, and definite claiming. The commencement of the opposition is conditioned upon the USPTO Director's determination that the opponent has raised a substantial question of patentability with respect to at least one claim in the patent.

H.R. 2795 further provides that opposition proceedings will be tried before a panel of three administrative patent judges. In the event that the patentee files an infringement suit within nine months of patent issuance, or six months of notifying the alleged infringer, the opposition will be stayed upon the request of the patent owner. The patent owner may amend its claims during an opposition, provided that those amendments do not broaden the scope of the claims. The opposition must conclude within one year of its commencement, although one six-month extension is possible. The results of opposition proceedings may be appealed to the courts.

Many observers have called for the United States to adopt an opposition system in order to provide more timely, lower cost, and more efficient review of issued patents.[190] Such a system could potentially improve the quality of issued patents by weeding out invalid claims. It might also encourage innovative firms to review issued patents soon after they are granted, thereby increasing the opportunity for technology spillovers.[191] Concerns have arisen over oppositions because they too may be costly, complex, and prone to abuse as a means for harassing patent owners.[192] A successful opposition proceeding would require a balancing of these concerns.

CONCLUDING OBSERVATIONS

Pending legislation would arguably work the most sweeping reforms to the U.S. patent system since the nineteenth century. However, many of these proposals, such as pre-issuance publication, prior user rights, and oppositions, have already been implemented in U.S. law to a more limited extent. These and other proposed modifications, such as the first-inventor-to-file priority system and elimination of the best mode requirement, also reflect the decades-old patent practices of Europe, Japan, and our other leading trading partners. As well, many of these changes enjoy the support of diverse institutions, including the Federal Trade Commission, National Academies, economists, industry representatives, attorneys, and legal academics.

Other knowledgeable observers are nonetheless concerned that certain of these proposals would weaken the patent right, thereby diminishing needed incentives for innovation. Some also believe that changes of this magnitude, occurring at the same time, do not present the most prudent course for the patent system. Patent reform therefore confronts Congress with difficult legal, practical, and policy issues, but also with the apparent possibility for altering and potentially improving the legal regime that has long been recognized as an engine of innovation within the U.S. economy.

REFERENCES

[1] Statistics from the United States Patent and Trademark Office (USPTO) support this account. In 1980, the USPTO received 104,329 utility patent applications; by 2003, this number had grown to 342,441 applications. During the same time period, the number of U.S. patents granted on an annual basis grew from 61,810 to 169,028. U.S. Patent and Trademark Office, *U.S. Patent Statistics, Calendar Years 1863 - 2003* [available at [http://www.uspto.gov]].

[2] National Research Council, National Academy of Sciences, *A Patent System for the 21st Century*, [Washington, National Academies Press, 2004] and Federal Trade Commission, *To Promote Innovation: The Proper Balance of Competition and Patent Law and Policy*, October 2003, available at [http://www.ftc.gov].

[3] P.L. 82-593, 66 Stat. 792 (codified at Title 35 United States Code).

[4] 35 U.S.C. § 111.

[5] 35 U.S.C. § 131.

[6] John R. Thomas, "On Preparatory Texts and Proprietary Technologies: The Place of Prosecution Histories in Patent Claim Interpretation," 47 *UCLA Law Review* (1999), 183.

[7] 35 U.S.C. § 112.

[8] Ibid.

[9] 35 U.S.C. § 101.

[10] 35 U.S.C. § 102.

[11] 35 U.S.C. § 103.

[12] 35 U.S.C. § 271(a).

[13] 35 U.S.C. § 283.

[14] 35 U.S.C. § 284.

[15] 35 U.S.C. § 154(a)(2). Although patent term is based upon the filing date, the patentee gains no enforceable legal rights until the USPTO allows the application to issue as a granted patent. A number of Patent Act provisions may modify the basic 20-year term, including examination delays at the USPTO and delays in obtaining marketing approval for the patented invention from other federal agencies.

[16] 35 U.S.C. § 282.

[17] 28 U.S.C. § 1295(a)(1).

[18] 28 U.S.C. § 1254(1).

[19] *See* Rebecca S. Eisenberg, "Patents and the Progress of Science: Exclusive Rights and Experimental Use," 56 *University of Chicago Law Review* 1017 (1989).

[20] 35 U.S.C. § 112.

[21] 35 U.S.C. § 154.

[22] Eisenberg, *supra,* at 1017.

[23] Robert P. Merges, "Intellectual Property and the Costs of Commercial Exchange: A Review Essay," 93 *Michigan Law Review* (1995), 1570.

[24] David D. Friedman *et al.*, "Some Economics of Trade Secret Law," 5 *Journal of Economic Perspectives* (1991), 61.

[25] *See* Frederic M. Sherer, *Industrial Market Structure and Economic Performance* (1970), 384-87.

[26] *See* John R. Thomas, "Collusion and Collective Action in the Patent System: A Proposal for Patent Bounties," *University of Illinois Law Review* (2001), 305.

[27] *Ibid.*

[28] *See* Dan L. Burk & Mark A. Lemley, "Is Patent Law Technology-Specific?," 17 *Berkeley Technology Law Journal* (2002), 1155.

[29] CRS Report RL31281, *Patent Quality and Public Policy: Issues for Innovative Firms in Domestic Markets*, by John R. Thomas.

[30] *See* Joseph Farrell & Robert P. Merges, "Incentives to Challenge and Defend Patents: Why Litigation Won't Reliably Fix Patent Office Errors and Why Administrative Patent Review Might Help," 19 *Berkeley Technology Law Journal* (2004), 943.

[31] *See* Bronwyn H. Hall & Dietmar Harhoff, "Post-Grant Reviews in the U.S. Patent System — Design Choices and Expected Impact," 19 *Berkeley Technology Law Journal* (2004), 989.

[32] *See* Robert P. Merges, "As Many As Six Impossible Patents Before Breakfast: Property Rights for Business Concepts and Patent System Reform," 14 *Berkeley Technology Law Journal* (1999), 577.

[33] *See* Craig Allen Nard, "Certainty, Fence Building and the Useful Arts," 74 *Indiana Law Journal* (1999), 759.

[34] Mark A. Lemley, "Rational Ignorance at the Patent Office," 95 *Northwestern University Law Review* (2001), 1495.

[35] Steven J. Elleman, "Problems in Patent Litigation: Mandatory Mediation May Provide Settlement and Solutions," 12 *Ohio State Journal on Dispute Resolution* (1997), 759.

[36] Dee Gill, "Defending Your Rights: Protecting Intellectual Property is Expensive," *Wall Street Journal* (25 Sep. 2000), 6.

[37] Mark H. Webbink, "A New Paradigm for Intellectual Property Rights in Software," 2005 *Duke Law and Technology Review* (May 1, 2005), 15.

[38] *See* Sarah Lai Stirland, "Will Congress Stop High-Tech Trolls?," *National Journal* (Feb. 26, 2005), 612.

[39] "Microsoft Advocates for Patent Reform," *eWEEK* (March 10, 2005).

[40] Stirland, *supra*, at 613.

[41] CRS Report RL31132, *Multinational Patent Acquisition and Enforcement: Public Policy Challenges and Opportunities for Innovative Firms*, by John R. Thomas.

[42] *See* Elizabeth D. Ferrill, "Patent Investment Trusts: Let's Build a Pit to Catch the Patent Trolls," 6 *North Carolina Journal of Law and Technology* (2005), 367.

[43] *See* Lorraine Woellert, "A Patent War Is Breaking Out on the Hill," *BusinessWeek* 45 (July 4, 2005).
[44] Nicholas Varchaver, "The Patent King," *Fortune* (May 14, 2001), 202.
[45] Michael Ravnitsky, "More Lemelson Suits," *The National Law Journal* (Dec. 17, 2001), B9.
[46] James Pooley, "Opinion: U.S. patent reform-a good invention," *Electronic Business* (1 Jan. 2000), 72.
[47] *See* Ronald J. Mann, "Do Patents Facilitate Financing in the Software Industry?," 83 *Texas Law Review* (2005), 961.
[48] *See* David G. Barker, "Troll or No Troll? Policing Patent Usage with An Open Post-Grant Review," 2005 *Duke Law and Technology Review* (Apr. 15, 2005), 11.
[49] 35 U.S.C. § 112.
[50] National Science Board, *Science and Engineering Indicators, 1993* (Dec. 8, 1993), 185. *See also* CRS Report RL30216, *Small, High Tech Companies and Their Role in the Economy: Issues in the Reauthorization of the Small Business Innovation (SBIR) Program*, by Wendy H. Schacht.
[51] For example, the National Academy of Engineering concluded that "small high-tech companies play a critical and diverse role in creating new products and services, in developing new industries, and in driving technological change and growth in the U.S. economy." National Academy of Engineering, *Risk & Innovation: The Role and Importance of Small High-Tech Companies in the U.S. Economy* (Washington: National Academy Press, 1995), 37. This assessment was founded on the ability of small firms to develop markets rapidly, generate new goods and services, and offer diverse products. The study also concluded that small businesses were less risk adverse than larger, established corporations and were often better positioned to exploit market opportunities quickly. A National Science Foundation report found that entrepreneurs and small firms are six times as effective as larger firms in utilizing research and development expenditures to generate new products. National Science Board, *Science and Engineering Indicators, 1993*, (Dec. 8, 1993), 185. Anderson, Anne, "Small Businesses Make it Big in the SBIR Program," *New Technology Week* (June 6, 1998), p. 2.
[52] Sally Wyatt & Gilles Y. Bertin, *Multinationals and Industrial Property* 139 (Harvester 1988).
[53] J. Douglas Hawkins, "Importance and Access of International Patent Protection for the Independent Inventor," 3 *University of Baltimore Intellectual Property Journal* (1995), 145.
[54] P.L. 96-517, 94 Stat. 2311 (codified at 35 U.S.C. §§ 200-212).
[55] CRS Report RL32076, *The Bayh-Dole Act: Selected Issues in Patent Policy and the Commercialization of Technology*, by Wendy H. Schacht.
[56] Josh Lerner, "Patent Policy Innovations: A Clinical Examination," 53 *Vanderbilt Law Review* (2000), 1841.
[57] *See* Arti K. Rai & Rebecca S. Eisenberg, "Bayh-Dole Reform and the Progress of Biomedicine," 66 *Law and Contemporary Problems* (Winter/Spring 2003), 289.
[58] 35 U.S.C. § 41(g).
[59] CRS Report RS20906, *U.S. Patent and Trademark Office Appropriations Process: A Brief Explanation*, by Wendy H. Schacht.

[60] Under a rule of "prior user rights," when a conflict exists between an issued patent and an earlier user of the patented technology, the validity of the patent is upheld but the prior user is exempted from infringement. *See* Pierre Jean Hubert, "The Prior User Right of H.R. 400: A Careful Balancing of Competing Interests," 14 *Santa Clara Computer and High Technology Law Journal* (1998), 189. Prior user rights are discussed further in this report below.

[61] *See* Gary L. Griswold & F. Andrew Ubel, "Prior User Rights — A Necessary Part of a First-to-File System," 26 *John Marshall Law Review* (1993), p. 567.

[62] *See* David H. Hollander, Jr., "The First Inventor Defense: A Limited Prior User Right Finds Its Way Into U.S. Patent Law," 30 *American Intellectual Property Law Association Quarterly Journal* (2002), 37 (noting the perception that prior user rights favor large, well-financed corporations).

[63] In particular, economic research suggests that different industries attach widely varying values to patents. For example, one study of the aircraft and semiconductor industries suggested that lead time and the strength of the learning curve were superior to patents in capturing the value of investments. In contrast, members of the drug and chemical industries attached a higher value to patents. Differences in the perception of the patent system have been attributed to the extent to which patents introduced significant duplication costs and times for competitors of the patentee. Richard C. Levin, Alvin K. Klevorick, Richard R. Nelson, and Sidney G. Winter, "Appropriating the Returns for Industrial Research and Development," Brookings Papers on Economic Activity, 1987, in *The Economics of Technical Change*, eds. Edwin Mansfield and Elizabeth Mansfield (Vermont, Edward Elgar Publishing Co., 1993), p. 254.

[64] Federic M. Scherer, "The Economics of Human Gene Patents", 77 *Academic Medicine* (Dec. 2002), p. 1350.

[65] Mann, *supra*, at 979.

[66] H.R. 2795, § 3.

[67] *See* Roger E. Schechter & John R. Thomas, *Principles of Patent Law* § 1.2.5 (2d ed. 2004).

[68] In addition, the party that was the first to invent must not have abandoned, suppressed or concealed the invention. 35 U.S.C. § 102(g)(2).

[69] *See* Charles E. Gholz, "First-to-File or First-to-Invent?", 82 *Journal of the Patent and Trademark Office Society* (2000), p. 891.

[70] *See* Peter A. Jackman, "Adoption of a First-to-File System: A Proposal," 26 *University of Baltimore Law Review* (1997), 67.

[71] 35 U.S.C. § 135.

[72] *See* Clifford A. Ulrich, "The Patent Systems Harmonization Act of 1992: Conformity at What Price?," 16 *New York Law School Journal of International and Comparative Law* (1996), p. 405.

[73] *See* Coe A. Bloomberg, "In Defense of the First-to-Invent Rule," 21 *American Intellectual Property Law Quarterly Journal* (1993), p. 255.

[74] *See* Bernarr A. Pravel, "Why the United States Should Adopt the First-to-File System for Patents," 22 *St. Mary's Law Journal* (1991), p. 797.

[75] 35 U.S.C. § 111(b).

[76] Gerald J. Mossinghoff, "The U.S. First-to-Invent System Has Provided No Advantage to Small Entities," 84 *Journal of the Patent and Trademark Office Society* (2002), p. 425.

[77] *See generally* Charles R.B. Marcedo, "First-to-File: Is American Adoption of the International Standard in Patent Law Worth the Price?," 18 *American Intellectual Property Law Association Quarterly Journal* (1990), p. 193.

[78] 35 U.S.C. § 102(g)(2).

[79] *See Del Mar Engineering Labs. v. United States*, 524 F.2d 1178 (Ct. Cl. 1975).

[80] 35 U.S.C. § 101.

[81] H.R. 2795, § 3(i).

[82] 35 U.S.C. § 102(b).

[83] Schechter & Thomas, *supra*, at § 4.3.1.

[84] European Patent Convention, Article 54(2).

[85] Japanese Patent Act, Article 29(1).

[86] 35 U.S.C. § 102(a).

[87] Convention of Paris for the Protection of Industrial Property, 13 U.S.T. 25 (1962).

[88] *See* G.H.C. Bodenhausen, *Guide to the Paris Convention for the Protection of Industrial Property* (United International Bureau for the Protection of Intellectual Property, Geneva, Switzerland 1968).

[89] 35 U.S.C. § 119(a).

[90] H.R. 2795, § 11(h).

[91] *See* Beedle v. Bennett, 122 U.S. 71 (1887).

[92] *See* Mendenhall v. Astec Indus., Inc., 13 USPQ2d 1913, 1937 (E.D. Tenn.1988), *aff'd*, 887 F.2d 1094 (Fed. Cir. 1989).

[93] *See generally* A.C. Auckerman & Co. v. R.L. Chaides.Construction Co., 960 F.2d 1020 (Fed. Cir. 1992).

[94] Schechter & Thomas, *supra*, at § 4.3.8.

[95] Bate Refrigerating Co. v. Sulzberger, 157 U.S. 1, 27 (1895) (quoting Ex parte Mushet, 1870 Comm'r Dec. 106, 108 (1870)).

[96] *See* Donald S. Chisum, "Foreign Activity: Its Effect on Patentability under United States Law," 11 *International Review of Industrial Property & Copyright Law* (1980), 26.

[97] *See* Schechter & Thomas, *supra*, at § 4.4.4.

[98] 35 U.S.C. § 111.

[99] 35 U.S.C. § 118.

[100] H.R. 2795, § 4(c).

[101] President's Commission on the Patent System,*"To Promote the Progress of . . . Useful Arts" in an Age of Exploding Technology* (1966).

[102] Advisory Commission on Patent Reform, *A Report to the Secretary of Commerce* (Aug. 1992), 179.

[103] *Ibid.*

[104] 35 U.S.C. § 112.

[105] *See, e.g., Chemcast Corp. v. Arco Industries Corp.* 913 F.2d 923 (Fed. Cir. 1990).

[106] *See* Jerry R. Selinger, "In Defense of the 'Best Mode': Preserving the Benefit of the Bargain for the Public, 43 *Catholic University Law Review* (1994), 1071.

[107] *See, e.g.,* Steven B. Walmsley, "Best Mode: A Plea to Repair or Sacrifice This Broken Requirement of United States Patent Law," 9 *Michigan Telecommunications and Technology Law Review* (2002), p. 125.
[108] 35 U.S.C. § 101.
[109] 1992 Advisory Commission Report, *supra,* at 102-03.
[110] *Ibid* at 101.
[111] *Ibid.* at 102-03.
[112] 35 U.S.C. § 122(a) (stating general rule that "applications for patents shall be kept in confidence by the Patent and Trademark Office and no information concerning the same given without authority of the applicant . . . ").
[113] Glaverbel Societe Anonyme v. Northlake Mktg. & Supply Inc., 45 F.3d 1550 (Fed. Cir. 1995).
[114] Heidelberger Druckmaschinen AG v. Hantscho Comm'l Prods., Inc., 21 F.3d 1068 (Fed. Cir. 1993).
[115] Jazz Photo Corp. v. U.S. Int'l Trade Comm'n, 264 F.3d 1094 (Fed. Cir. 2001).
[116] John F. Lynch, "An Argument for Eliminating the Defense of Patent Unenforceability Based on Inequitable Conduct," 16 *American Intellectual Property Law Association Quarterly Journal* (1988), 7.
[117] Burlington Indus., Inc. v. Dayco Corp., 849 F.2d 1418 (Fed. Cir. 1988).
[118] *See, e.g.,* Scott D. Anderson, "Inequitable Conduct: Persistent Problems and Recommended Resolutions," 82 *Marquette Law Review* (1999), 845.
[119] Lynch, *supra*, at 7.
[120] 1992 Advisory Commission, *supra*, at 114.
[121] H.R. 2795, § 5(a).
[122] 37 C.F.R. § 1.56.
[123] R. Carl Moy, "The Effect of New Rule 56 on the Law of Inequitable Conduct," 74 *Journal of the Patent and Trademark Office Society* (1992), 391.
[124] *See* Harry F. Manbeck, Jr., "The Evolution and Issue of New Rule 56," 20 *American Intellectual Property Law Association Quarterly Journal* (1992), 136.
[125] H.R. 2795, § 5
[126] Schechter & Thomas, *supra*, at § 7.2.6.
[127] John C. Todaro, "Potential Upcoming Changes in U.S. Patent Laws: the Publication of Patent Applications," 36 *IDEA: Journal of Law and Technology* (1996), 309.
[128] 35 U.S.C. § 122(b).
[129] H.R. 2795, § 9(a).
[130] 37 C.F.R. § 1.291.
[131] 35 U.S.C. § 122(c).
[132] 37 C.F.R. § 1.99.
[133] U.S. Dept. of Commerce, U.S. Patent & Trademark Off., Manual of Patent Examining Procedure § 1134.01 (8th ed. May 2004).
[134] H.R. 2795, § 10.
[135] 35 U.S.C. § 120.
[136] Schechter & Thomas, *supra*, at § 7.2.4.
[137] Mark A. Lemley & Kimberly A. Moore, "Ending Abuse of Patent Continuations," 84 *Boston University Law Review* (2004), 63.
[138] 35 U.S.C. § 122(b)(1).

[139] 35 U.S.C. § 122(b)(2).
[140] 35 U.S.C. § 154(a)(2).
[141] H.R. 2795, § 8.
[142] Restatement of Unfair Competition § 39.
[143] David D. Friedman, "Some Economics of Trade Secret Law," 5 *Journal of Economic Perspectives* (1991), 61, 64.
[144] 35 U.S.C. § 102(b). *See* Metallizing Engineering Co. v. Kenyon Bearing & Auto Parts, 153 F.2d 516 (2d Cir. 1946).
[145] 35 U.S.C. § 154.
[146] W.L. Gore & Associates. v. Garlock, Inc., 721 F.2d 1540 (Fed. Cir. 1983).
[147] 149 F.3d 1368 (Fed. Cir. 1998).
[148] As presently codified at 35 U.S.C. § 273, the first inventor defense is subject to a number of additional qualifications. First, the defendant must have reduced the infringing subject matter to practice at least one year before the effective filing date of the application. Second, the defendant must have commercially used the infringing subject matter prior to the effective filing date of the patent. Finally, any reduction to practice or use must have been made in good faith, without derivation from the patentee or persons in privity with the patentee.
[149] The bill would also remove the requirement that the prior use be reduced to practice at least one year before the effective filing date of such patent. Under H.R. 2795, the defense would apply where reduction to practice occurred prior to the patent's filing date.
[150] *See* Gary L. Griswold & F. Andrew Ubel, "Prior User Rights — A Necessary Part of a First-to-File System," 26 *John Marshall Law Review* (1993), 567.
[151] Paul R. Morico, "Are Prior User Rights Consistent with Federal Patent Policy?: The U.S. Considers Legislation to Adopt Prior User Rights," 78 *Journal of the Patent and Trademark Office Society* (1996), 572.
[152] *See* Robert L. Rohrback, "Prior User Rights: Roses or Thorns?," 2 *University of Baltimore Intellectual Property Review* (1993), 1.
[153] 35 U.S.C. § 283.
[154] H.R. 2795, § 7.
[155] *See* Richardson v. Suzuki Motor Co., 868 F.2d 1226, 1247 (Fed. Cir. 1989).
[156] MercExchange, L.L.C. v. eBay, Inc., 401 F.3d 1323, 1338 (Fed. Cir. 2005) (select citations omitted).
[157] 69 F.2d 577 (7th Cir. 1934).
[158] *Ibid.* at 593.
[159] 146 F.2d 941 (9th Cir. 1944).
[160] *Ibid.* at 943-44.
[161] 852 F. Supp. 813 (D. Minn.1994).
[162] *Ibid.* at 850 — 51.
[163] *Ibid.* at 862.
[164] 35 U.S.C. § 283.
[165] W. David Westergard, Remedying the Growing Abuse of the Patent System Through Targeted Legislation, *in Conference Papers on International Intellectual Property Law & Policy* (Fordham Law School March 31, 2005).

[166] *See* Lorraine Woellert, "A Patent War Is Breaking Out on the Hill," *BusinessWeek* 45 (July 4, 2005).
[167] Rite-Hite Corp. v. Kelley Co., Inc., 56 F.3d 1538, 1547 (Fed. Cir. 1995).
[168] *See, e.g.,* Frank H. Easterbrook, "Intellectual Property is Still Property," 13 *Harvard Journal of Law and Public Policy* (1990), 108.
[169] *See* Mark A. Lemley, "Property, Intellectual Property, and Free Riding," 83 *Texas Law Review* (2005), 1031.
[170] *See* Michael A. Carrier, "Cabining Intellectual Property Through a Property Paradigm," 54 *Duke Law Journal* (2004), 1.
[171] H.R. 2795, § 6.
[172] Westergard, *supra*, at 7.
[173] Stratoflex, Inc. v. Aeroquip Corp., 713 F.2d 1530, 1540 (Fed. Cir. 1983).
[174] Allen Eng'g Corp. v. Bartell Indus., Inc., 299 F.3d 1336, 1345 (Fed. Cir. 2002).
[175] 35 U.S.C. § 284.
[176] *Ibid.*
[177] *See* Read Corp. v. Portec, Inc., 970 F.2d 816, 826 (Fed. Cir. 1992).
[178] *See* Beatrice Foods Co. v. New England Printing & Lithographing Co., 923 F.2d 1576, 1578 (Fed. Cir.1991).
[179] *See, e.g.,* Jon E. Wright, "Willful Patent Infringement and Enhanced Damages — Evolution and Analysis," 10 *George Mason Law Review* (2001), 97.
[180] *See, e.g.,* Fromson v. Western Litho Plate & Supply Co., 853 F.2d 1568, 1572 (Fed. Cir.1988).
[181] 383 F.3d 1337 (Fed. Cir. 2004).
[182] *Ibid.* at 1341.
[183] *See generally* Schechter & Thomas, *supra*, at § 9.2.5.
[184] H.R. 2795, § 6.
[185] Mark D. Janis, "Inter Partes Reexamination," 10 *Fordham Intellectual Property, Media & Entertainment Law Journal* (2000), 481.
[186] Craig Allen Nard, "Certainty, Fence Building and the Useful Arts," 74 *Indiana Law Journal* (1999), 759.
[187] *See* Schechter & Thomas, *supra*, at § 7.5.4.
[188] 35 U.S.C. § 315(c).
[189] H.R. 2795, § 9(d).
[190] *See* National Research Council of the National Academies, *A Patent System for the 21st Century* (2004), 96.
[191] *Ibid.* at 103.
[192] *See* Mark D. Janis, "Rethinking Reexamination: Toward a Viable Administrative Revocation System for U.S. Patent Law," 11 *Harvard Journal of Law and Technology* (1997), 1.

In: New Intellectual Property Issues
Editor: John V. Luong, pp. 37-56
ISBN: 978-1-60021-211-6
© 2007 Nova Science Publishers, Inc.

Chapter 2

SCIENTIFIC RESEARCH AND THE EXPERIMENTAL USE PRIVILEGE IN PATENT LAW*

John R. Thomas

SUMMARY

Congress has identified research and development (R&D) as important contributors to technological progress. The performance of R&D may have intellectual property ramifications, however. To the extent that researchers use patented inventions without authorization, they may face infringement liability. Although the courts recognize an exception to patent infringement known as the "experimental use privilege," this judicially created doctrine has been described as very narrow and rarely applied. In particular, the experimental use privilege applies only to uses done for amusement, to satisfy idle curiosity or for strictly philosophical inquiry. This doctrine does not excuse uses that are in keeping with the accused infringer's business objectives.

In 2002, the U.S. Court of Appeals for the Federal Circuit applied these principles in the case of *Madey v. Duke University*. The court held that the experimental use privilege does not apply to activities that are "in keeping with the alleged infringer's legitimate business" — even though the business of the defendant, Duke University, was nonprofit research. This ruling has raised concerns among some representatives of universities and research institutions, who fear that their basic R&D activities will subject them to patent infringement lawsuits.

Competing views have arisen over the significance of the *Madey v. Duke University* case. Some commentators believe that a limited experimental use privilege may best encourage technological advancement by rewarding successful researchers with robust patent rights. Others argue that the restricted nature of the experimental use privilege may in fact limit researcher access to state-of-the-art technologies and thus discourage further technological development. Still others assert that this issue is not of great practical importance, as few patent owners will likely file costly and time-consuming lawsuits against researchers who are not making commercially important uses of patented inventions.

The judicially created, "common law" experimental use privilege is complemented by a limited statutory experimental use privilege for patents on pharmaceuticals, medical devices, and certain other products regulated by the Food and Drug Administration. This provision, enacted as part of the 1984 Hatch-Waxman Act, applies to firms seeking to market generic

* Excerpted from CRS Report RL32651, Updated October 28, 2004

equivalents of brand-name products. In addition, Congress has enacted other intellectual property legislation that incorporates provisions shielding researchers from infringement liability.

Should congressional interest continue in this area, a variety of options are available. If the current scope of the common law experimental use privilege is deemed to be appropriate, then no action need be taken. Alternatively, Congress could enact legislation confirming the limited experimental use privilege recognized in *Madey v. Duke University* and predecessor cases. Introduction of a broader form of the experimental use privilege into U.S. patent law is an additional possibility.

Congress has identified research and development (R&D) as important contributors to technological advance.[1] Technological improvements are in turn acknowledged to be a driving force in the long-term economic expansion of the United States. The federal government plays a role in funding R&D in the United States, having appropriated an estimated $126 billion for federal R&D in 2004.[2] Legislation has further encouraged technical cooperation among government, academia, and the private sector during their R&D efforts.[3]

Although the issue has not commonly arisen, R&D activities may have intellectual property implications. To the extent that researchers use another's patented invention without authorization, they may face liability for patent infringement. While the courts recognize an exception to patent infringement known as the "experimental use privilege,"[4] this judicially created doctrine has been described as "very narrow"[5] and "only sparingly applied."[6] The experimental use privilege allows a researcher to use a patented invention without permission from or compensation to the patent owner. However, such a use must be "merely for philosophical experiments, or for the purpose of ascertaining the sufficiency of the [patented invention] to produce its described effects."[7] If the purpose of the research falls outside these parameters, the "common law" experimental use privilege is inapplicable, and the researcher may face liability for patent infringement.

Congress has supplemented this "common law" experimental use privilege with a statutory research exemption to patent infringement. Introduced as part of the Hatch-Waxman Act,[8] this exemption is limited to patented pharmaceuticals, medical devices and certain other products regulated by the Food and Drug Administration ("FDA"). It exempts from patent infringement clinical studies and other activities performed in order to obtain FDA marketing approval.

Despite relatively limited judicial application of patent law's experimental use privilege, vigorous debate on the topic has persisted for many years.[9] The 2002 judicial opinion in *Madey v. Duke University* reinvigorated this discussion.[10] Here, the U.S. Court of Appeals for the Federal Circuit ("Federal Circuit") rejected an experimental use defense to a patent infringement lawsuit against Duke University. The Federal Circuit held that the experimental use privilege does not apply to infringing activities that are "in keeping with the alleged infringer's legitimate business" — even though Duke University's "business" consisted of serving as a nonprofit research institution. Although the experimental use privilege was not formally eliminated in *Madey v. Duke University*, some commentators believe that this infringement exception is so limited that it is practically nonexistent.[11] This decision has in turn raised concerns among some representatives of universities and research institutions, who fear that their basic research activities will subject them to patent infringement lawsuits.[12]

Both supporters and detractors of an expansive experimental use privilege believe that this doctrine bears upon the fundamental goal of the patent system, the promotion of innovation. A more narrow exception will result in broader intellectual property rights. Robust patent rights potentially provide innovators with strong incentives to invest in research and development.[13] On the other hand, an overly restrictive experimental use exception might conceivably depress technological advancement by decreasing the ability of researchers to experiment with state-of-the-art technology.[14] Debate over the appropriate scope of the experimental use exception has been particularly vigorous with respect to patents on cell lines, assays, and other "research tools" used primarily for the purpose of conducting biotechnology research.[15]

This report will explore the intellectual property laws and policies concerning the patent law's experimental use privilege. It begins with a review of patent policy and procedures. The report then discusses current statutes and judicial precedent governing the common law experimental use in patent law, as well as the potential impact of this doctrine upon innovation policy. The report next reviews the statutory experimental use privilege with respect to the Hatch-Waxman Act and other, more specialized intellectual property statutes. This report closes with a summary of congressional issues and alternatives.

INTRODUCTION TO THE PATENT SYSTEM

Patent Policy

The goal of the patent system is to promote the production and dissemination of technological information. Some observers believe that absent a patent system, individuals and firms would be less likely to engage in research and development. Without the availability of patent protection, new inventions could be easily copied by "free riders" who incurred no cost to develop and perfect the technology involved, and who would therefore be able to undersell the original inventor. The resulting inability of inventors to capitalize on their inventions would lead to an environment where too few inventions are made. By providing individuals with exclusive rights in their inventions for a limited time, the patent system allows inventors to realize the profits from their inventions.[16]

Other commentators believe that if the patent system were unavailable, individuals would maintain their inventions as trade secrets so that competitors could not exploit them. Trade secrets do not enrich the collective knowledge of society, however, nor do they discourage others from engaging in duplicative research. The patent system avoids these inefficiencies by requiring inventors to consent to the disclosure of their inventions in issued patent instruments.[17]

Additional explanations for the patent laws have been offered. The Patent Act may stimulate technological advancement by inducing individuals to "invent around" patented technology. Issued patent instruments may point the way for others to develop improvements, exploit new markets or discover new applications for the patented technology.[18] Moreover, the patent system may encourage patentees to commercialize their proprietary technologies during the term of the patent. The protection provided by a patent's proprietary rights increases the likelihood a firm will continue to refine, produce and market the patented technology.[19] Finally, the patent law has been identified as a facilitator of markets. Absent patent rights, an inventor may have scant tangible assets to sell or license, and even less ability to police the conduct of a contracting party. By reducing a licensee's opportunistic possibilities, the patent system lowers transaction costs and makes technology-based transactions more feasible.[20]

The patent system has inspired numerous critics, however. Some detractors have asserted that the patent system is unnecessary due to market forces that already suffice to create an optimal level of invention. The desire to gain a lead time advantage over competitors, as well as the recognition that technologically backward firms lose out to their rivals, may well provide sufficient inducement to invent without the need for further incentives.[21] Others observe that the inventions that fueled many dynamic sectors of modern industry, such as biotechnologies and computer software, arose at a time when patent rights were unavailable or uncertain.[22]

While these justifications and criticisms have varying degrees of intuitive appeal, none of them has been empirically validated. No authoritative study conclusively demonstrates that society obtains more rapid technological development with patents than it would without them. As a result, the rationales for, and criticisms of, the patent system remain open to challenge.[23]

Patent Acquisition and Enforcement

As mandated by the Patent Act of 1952,[24] U.S. patent rights do not arise automatically. Inventors must prepare and submit applications to the U.S. Patent and Trademark Office ("USPTO") if they wish to obtain patent protection.[25] USPTO officials, known as examiners, then assess whether the application merits the award of a patent.[26] The patent acquisition process is commonly known as "prosecution."[27]

In deciding whether to approve a patent application, a USPTO examiner will consider whether the submitted application fully discloses and distinctly claims the invention.[28] In addition, the application must disclose the "best mode," or preferred way, that the applicant knows to practice the invention.[29] The examiner will also determine whether the invention itself fulfills certain substantive standards set by the patent statute. To be patentable, an invention must be useful, novel and nonobvious. The requirement of usefulness, or utility, is satisfied if the invention is operable and provides a tangible benefit.[30] To be judged novel, the invention must not be fully anticipated by a prior patent, publication or other knowledge within the public domain.[31] A nonobvious invention must not have been readily within the ordinary skills of a competent artisan at the time the invention was made.[32]

If the USPTO allows the patent to issue, the patent proprietor obtains the right to exclude others from making, using, selling, offering to sell or importing into the United States the patented invention.[33] The term of the patent is ordinarily set at twenty years from the date the patent application was filed.[34] Patent title therefore provides inventors with limited periods of exclusivity in which they may practice their inventions, or license others to do so. The grant of a patent permits the inventor to receive a return on the expenditure of resources leading to the discovery, often by charging a higher price than would prevail in a competitive market.

Patent rights are not self-enforcing. A patentee bears responsibility for monitoring its competitors to determine whether they are using the patented invention or not. Patent proprietors who wish to compel others to observe their intellectual property rights must usually commence litigation in the federal district courts.[35] The U.S. Court of Appeals for the Federal Circuit possesses exclusive national jurisdiction over all patent appeals from the district courts,[36] while the U.S. Supreme Court possesses discretionary authority to review cases decided by the Federal Circuit.[37]

THE COMMON LAW EXPERIMENTAL USE PRIVILEGE

General Principles

Under the Patent Act of 1952, any individual who makes, uses, sells, offers to sell, or imports into the United States a patented invention without the authorization of the patent owner faces liability for infringement.[38] Although the Patent Act authorizes a number of defenses to a charge of patent infringement — such as that the patented invention does not meet the statutory standards for patentability and was improvidently awarded a patent by the USPTO[39] — the statute does not expressly authorize a generally applicable experimental use privilege. The federal courts have nonetheless developed a "common law" experimental use privilege using their judicial powers.

Judicial decisions from the nineteenth century established the scope of the experimental use privilege. In the first of these opinions, the 1813 case of *Whittemore v. Cutter*,[40] Justice Joseph Story explained that "it could never have been the intention of the legislature to punish a man, who constructed such a machine merely for philosophical experiments, or for the purpose of ascertaining the sufficiency of the machine to produce its described effects."[41] Later, in *Sawin v. Guild*,[42] Justice Story explained that an unauthorized manufacture of a patented invention was not an infringement unless it constituted "making with an intent to use for profit, and not for the mere purpose of philosophical experiment, or to ascertain the verity and exactness of the specification." The 1861 decision in *Peppenhausen v. Falke* further explained:

> It has been held, and no doubt is now well settled, that an experiment with a patented article for the sole purpose of gratifying a philosophical taste, or curiosity, or for mere amusement, is not an infringement of the rights of the patentee.[43]

This judicial conception of the experimental use privilege has been described as "crabbed,"[44] "narrowly construed"[45] and "rarely sustained."[46] As explained in numerous judicial opinions, the experimental use exception applies only to uses done for amusement, to satisfy idle curiosity or for strictly philosophical inquiry.[47] Any use that is commercial in nature is not subject to the doctrine.[48]

The opinion of the Court of Claims in *Pitcairn v. United States* provides one example of judicial views concerning the experimental use privilege.[49] In that case, the U.S. government was accused of infringing 59 patents relating to helicopters. The government contended that various aircraft used only for purposes of testing and demonstration should be held not to infringe. The court disagreed, explaining:

> Defendant urges the court to exclude from compensation any aircraft used by the defendant for testing, evaluational, demonstrational or experimental purposes. Use for such purposes is use by or for the Government and is compensable. Obviously every new helicopter must be tested for lifting ability, for the effect of vibration on installed equipment, flight speed and range, engine efficiency, and numerous other factors. Tests, demonstrations, and experiments of such nature are intended uses of the infringing aircraft manufactured for the defendant and are in keeping with the legitimate business of the using agency. Experimental use is not a defense in the present litigation.[50]

This language further suggests that if the use furthers the user's "legitimate business" objectives, even in a tangential way, then it is not experimental.[51]

A 2002 decision on the experimental use privilege, *Madey v. Duke University*,[52] reflects this narrow sense of the experimental use privilege. Here, Duke University recruited Dr. John M.J. Madey from Stanford University in order to serve as a research professor and director of a laser laboratory. A dispute ultimately led to Madey's resignation from Duke. After he left the university, Madey brought suit for infringement of two patents relating to the operation of specialized equipment used in the Duke laser laboratory.[53]

Responding to Madey's charges of patent infringement, Duke sought to take advantage of the experimental use privilege. Duke explained that it was a non-profit educational institution dedicated to teaching, research and the advancement of knowledge. Duke further argued that it does not undertake research or development work principally for the purpose of obtaining patents or designing products for the marketplace. The district court agreed that the experimental use privilege applied and ruled in favor of Duke.[54]

On appeal, the Federal Circuit reversed the district court's experimental use determination. The Court of Appeals characterized the experimental use privilege as "very narrow and strictly limited."[55] In particular, the Federal Circuit observed that the experimental use privilege "does not immunize any conduct that is in keeping with the alleged infringer's legitimate business, regardless of commercial implications."[56] It further explained that:

> Major research universities, such as Duke, often sanction and fund research projects with arguably no commercial application whatsoever. However, these projects unmistakably further the institution's legitimate business objectives, including educating and enlightening students and faculty participating in these projects. The projects also serve, for example, to increase the status of the institution and lure lucrative research grants, students and faculty.[57]

As a result, the Federal Circuit held that the district court had resolved the case based upon an inappropriately expansive view of the experimental use privilege. It remanded the litigation back to the district court, with instructions to resolve the issue in light of the Federal Circuit's decision.[58]

Many patent law experts agree that following the *Madey v. Duke University* case, colleges, universities and other academic institutions are unlikely to be able to rely upon the common law experimental use privilege as a defense to a charge of patent infringement.[59] Co-authors Lawrence Sung, then a member of the University of Maryland Law School faculty, and attorney Claire M. Maisano explained that the "decision in *Madey* leaves grave doubt that the common law exemption to patent infringement liability can act as a safe harbor for any academic research effort."[60] Commentator Michelle Cai further opined that "practically any project conducted by a research university, even one without any commercial implications, would be in keeping with the university's legitimate business interests and hence would not qualify for the experimental use defense."[61]

Observers generally agree that *Madey v. Duke University* either retains, or perhaps restricts to an even greater degree, the quite limited nature of the common law experimental use privilege as it might be applied outside of academic settings.[62] Attorneys Paul Devinksy and Mark G. Davis concluded that the opinion is consistent with previous judicial interpretations of the common law experimental use privilege. As a result the privilege "lives on as a narrow defense to a claim of infringement."[63] Other observers would go further. Ms. Cai states, for example, that the decision "has essentially destroyed any practical meaning to the experimental use defense."[64] Attorneys Cathryn Campbell and R.V. Lupo are in accord, stating that after *Madey v. Duke University* "the Experimental Use Exception would appear to provide little, if any protection in today's world."[65]

Experimental Use and Innovation Policy

Although the experimental use privilege has been part of the patent law for many years, the *Madey v. Duke University* opinion has renewed dialogue over the propriety and scope of this infringement exemption. Proponents and detractors of a broad experimental use privilege, as well as those who do not believe the doctrine is important as a practical matter, have expressed diverse opinions. This report summarizes these competing views.

Some commentators believe that a broad experimental use privilege is inappropriate as a matter of technology policy. Under this view, a liberal experimental allowance would greatly ease the ability of competitors to "design around" the invention or develop competing technologies. Patent owners in turn would be less able to appropriate the returns of their investments in research and development, this account continues, and would therefore be discouraged from making future investments in research and development.[66] As attorney Jordan Karp concludes: "Rather than spurring increased innovative activity, a broad experimental use exception would have just the opposite effect."[67]

For some observers, a broad experimental use privilege is inappropriate even when research takes place within an academic research setting. This is because university research is often not isolated from the private sector, but instead may have significant commercial implications.[68] As stated by two senior officers of the USPTO, the *Madey v. Duke University* decision "recognized a basic economic truth underlying research performed by large universities — it is a business, and universities derive substantial commercial value from that research."[69]

Indeed, some commentators believe that university research is increasingly likely to have commercial implications. This shift is believed to be due in part to federal legislation commonly known as the Bayh-Dole Act.[70] The Bayh-Dole Act aims to encourage the commercialization of basic research by allowing universities and small businesses to procure patents on inventions that result from federally funded research. Since the passage of the Bayh-Dole Act, many research universities have developed patent portfolios and garnered significant royalties from intellectual property licensing.[71] Because academic institutions are increasingly benefitted from the patent system, some observers reason, they should also be held accountable when they infringe the patents of others.[72]

In contrast, others believe that the common law experimental use privilege is overly narrow. They assert that the current scope of this doctrine too greatly restricts the ability of innovators to "tinker" with the developments of others.[73] Under this view, research may be chilled if scientists cannot experiment upon state-of-the-art technology free from charges of patent infringement. By limiting the tools with which researchers can work, these commentators say, the patent system could ultimately depress, rather than promote innovation.[74]

Other observers believe that limiting the experimental use privilege with respect to universities and nonprofit institutes could impede academic research. Some university-based scientists believe that, unlike some of their counterparts in the private sector, academic researchers have all but ignored the patent system.[75] In order to avoid patent infringement, universities may have to devote scarce resources to perform costly patent searches and engage in licensing negotiations with patent holders. To a greater extent than profit-seeking firms, educational institutions may find that these obligations weigh heavily on their frequently tight budgets.[76]

Several legal scholars have suggested that the holding of *Madey v. Duke University* raises particular concerns with regard to patented research tools. A "research tool" is an invention, such as a particular cell line, reagent, or antibody, that is used exclusively or primarily for the purpose of conducting scientific research.[77] Some observers believe that patents are too frequently granted on research tools, particularly in biotechnology, and that such patents may impede future advancement. For example, two members of the University of Michigan Law School faculty, Michael A. Heller and Rebecca S. Eisenberg, have concluded that researchers might someday need to obtain numerous patent licenses in order to conduct basic research.[78] The costs and complications of engaging in numerous patent transactions may potentially create an "anti-commons": an environment where resources that could be committed towards further research and development are inefficiently underutilized.[79] Under this view, the narrowly cabined experimental use privilege contemplated in *Madey v. Duke University* strengthens patents on research tools and might make research and development more difficult to accomplish.[80]

Notably, other scholars contest these theories. For example, F. Scott Kieff, a member of the law faculty of Washington University in St. Louis, asserts that this scenario is incorrect as a matter of individual incentives.[81] Rational patent holders should always encourage others to research with their technologies, Kieff explains, so as to increase the number of applications for their inventions and hence their own profits. Wesley M. Cohen, an academic at the Fuqua School of Business at Duke University, has also conducted empirical research suggesting that although the "anti-commons" environment posited by Heller and Eisenberg may be theoretically possible, to date it has not actually occurred.[82]

Other commentators believe that "scientists recognize the benefits of sharing materials freely whenever possible and have developed informal norms to achieve broad dissemination of research tools."[83] For example, the National Institutes of Health ("NIH") has issued "Principles and Guidelines" in order to promote the broad use of patented research tools in biotechnology. [84] In particular, the NIH encourages patent proprietors to license their proprietary research tools in such a way as to minimize restrictions upon their use. Although the Principles and Guidelines formally apply only to recipients of federal funding, the NIH has urged the entire biotechnology community to adopt similar policies "so that all biomedical research and development can be synergistic and accelerated."[85]

Finally, a third set of commentators remain agnostic about the propriety of an experimental use privilege, but believe that this issue is not of great importance for practical reasons. Patent infringement litigation is widely regarded as costly, time-consuming and complex.[86] There may be insufficient economic justification to commence litigation against individuals who are not making commercially important uses of patented inventions. As a result, patent infringement suits may only rarely be brought against hobbyists, philosophers and noncommercial defendants, regardless of how narrowly or broadly the experimental use privilege is defined. [87]

It is also important to remember that the patent law may, in certain circumstances, provide researchers with the ability to use products even though they have been patented by others. One of these principles is known as the "exhaustion" doctrine. Under this legal rule, once a patent owner has sold a patented product, he cannot control the use of that particular product. Any patent rights in that specific physical item are said to have been "exhausted" by this initial sale. Sometimes the exhaustion principle is termed the "first sale" doctrine.[88]

For example, suppose that a pharmaceutical firm wished to analyze a drug that had been patented by another. That firm might wish to confirm the drug's biological activity, identify new medical indications, or compare its pharmacological profile to those of other compounds. If the firm is able simply to purchase the patented drug on the open market, then no issues of patent infringement will ordinarily arise. Any patent on the drug is exhausted once the drug has been sold, allowing the purchasing firm to use the patented drug as it wishes. The scope of the experimental use privilege is irrelevant in this scenario.[89]

On the other hand, suppose that a patented drug is not available for purchase within the market. In order to experiment with that compound, a pharmaceutical firm must synthesize it within its own laboratories. This step would be an act of patent infringement, however, because the right to make a patented invention is exclusive to the patent owner.[90] In this case the experimental use privilege would, at least theoretically, come into play as a possible defense to patent infringement.

Statutory Experimental Use Privileges

The existence, possible scope and importance of the common law experimental use privilege in patent law remains the subject of considerable debate. However, it should be noted that the Patent Act of 1952 includes a limited statutory experimental use privilege for patents on pharmaceuticals, medical devices, and certain other products regulated by the Food and Drug Administration (FDA). This provision, enacted as part of 1984 legislation known as the Hatch-Waxman Act,[91] applies to firms seeking to market generic equivalents of brand-name products. In addition, Congress has enacted other intellectual property legislation that incorporates provisions shielding researchers from infringement liability. This report considers these topics in turn.

The Hatch-Waxman Act

The Hatch-Waxman Act for the first time introduced a statutory experimental use privilege into the patent laws. This privilege applies only to certain products —notably specific kinds of pharmaceuticals and medical devices — that are regulated by the FDA. Firms must obtain FDA approval in order to market these products. Ordinarily the FDA will approve only of those products that have been proven to be safe and effective through laboratory, animal and clinical investigations.[92] Such studies can be costly and time-consuming. In some cases, the effort to obtain FDA marketing approval requires many years to complete.[93]

Many firms wish to sell drugs or medical devices that are "generic" — that is to say, equivalent to a product that was first developed and sold under a brand name by a different company. Prior to Hatch-Waxman Act, generic firms faced two notable difficulties in getting their products to market. First, generic firms ordinarily had to conduct the same sort of expensive and lengthy clinical investigations as their brand-name counterparts in order to obtain FDA marketing approval. This requirement existed even in circumstances where the generic drug was chemically identical to a brand-name drug of widely acknowledged safety and effectiveness.[94]

Second, generic firms had to account for the patents owned by their brand-name competitors. As of the early 1980's, legal uncertainty existed as to whether a generic firm could conduct clinical trials at all if a brand-name firm held patents on the drug or medical device. As part of the FDA marketing approval process, the generic firm would need to both make and use the patented drug or device —activities that under the intellectual property laws are exclusive to the patent proprietor.[95] The result was that a generic firm could be sued for patent infringement and, at least until the relevant patents expired, enjoined from engaging in the activities it needed to perform in order to satisfy FDA marketing approval requirements.

Some generic drug and medical device firms sought to rely upon the common law experimental use privilege in this situation. Their position was that activities performed in order to fulfill FDA marketing approval standards were merely experimental in nature, and as a result should be exempted from patent infringement.[96] Whether the courts would uphold this argument remained an open legal question for many years.

Eventually this issue came before the Federal Circuit, which in 1984 issued its decision in *Roche Products, Inc. v. Bolar Pharmaceutical Co.*[97] Here the Federal Circuit conclusively held that the common law experimental use privilege did not shield generic firms engaged in FDA marketing approval activities from charges of patent infringement. In this case, Roche Products, Inc. ("Roche") marketed a prescription sleeping pill under the trademark "Dalmane." Roche also was the proprietor of a patent claiming a chemical compound, flurazepam hcl, that was the active ingredient in Dalmane. Bolar Pharmaceutical Co. ("Bolar"), a manufacturer of generic drugs, grew interested in marketing a generic equivalent of Dalmane. Prior to the expiration of Roche's patent, Bolar obtained a supply of flurazepam hcl from a foreign manufacturer. It began to form the flurazepam hcl into dosage form capsules to obtain stability data, dissolution rates, bioequivalency studies and blood serum studies necessary to obtain marketing approval from the FDA.[98]

The Federal Circuit concluded Bolar's activities infringed the Roche patents, and that the experimental use defense did not apply. The Court of Appeals reasoned:

> Bolar's intended "experimental" use is solely for business reasons and not for amusement, to satisfy idle curiosity, or for strictly philosophical inquiry. Bolar's intended use of flurazepam hcl to derive FDA required test data is thus an infringement of the [Roche] patent. Bolar may intend to perform "experiments," but unlicensed experiments conducted with a view to the

> adaptation of the patented invention to the experimentor's business is a violation of the rights of the patentee to exclude others from using his patented invention. It is obvious here that it is a misnomer to call the intended use de minimus. It is no trifle in its economic effect on the parties even if the quantity used is small. It is not a dilettante affair such as Justice Story envisioned. We cannot construe the experimental use rule so broadly as to allow a violation of the patent laws in the guise of "scientific inquiry," when that inquiry has definite, cognizable, and not insubstantial commercial purposes.[99]

Congress responded to *Roche v. Bolar* by enacting the statute commonly known as the Hatch — Waxman Act. The statute introduced a number of changes to both the patent law and the food and drug law.[100] Among them was an accelerated marketing approval process for generic products.[101] In addition, the Hatch-Waxman Act created a statutory exemption from patent infringement for activities associated with regulatory marketing approval. As originally enacted and codified in 35 U.S.C. § 271(e)(1), the Hatch-Waxman Act exempted from patent infringement "uses reasonably related to the development and submission of information under a Federal law which regulates the manufacture, use, or sale of drugs." Through the Generic Animal Drug and Patent Term Restoration Act, which became effective on November 16, 1988, Congress extended this provision to cover regulated veterinary drugs and biological products as well.[102]

A number of significant judicial opinions have interpreted the Hatch-Waxman Act's experimental use privilege. The U.S. Supreme Court opinion in *Eli Lilly and Co. v. Medtronic* clarified the sorts of products that are covered by this statute.[103] In that case, Eli Lilly, which owned a patent claiming a cardiac defibrillator, filed an infringement action against Medtronic for its use of a similar device. Medtronic in turn pointed to 35 U.S.C. § 271(e)(1), arguing that Medtronic's use of the defibrillator was reasonably related to obtaining data for FDA approval. Eli Lilly in turn argued that the wording of 35 U.S.C. § 271(e)(1) — as it was at the time — only expressly referred to "drugs," not "medical devices."[104]

The Supreme Court sided with Medtronic, concluding that all of the products eligible for patent term extension under the Hatch-Waxman Act fall within the scope of the 35 U.S.C. § 271(e)(1) experimental use privilege. Justice Scalia concluded that the statutory phrase "a Federal law which regulates the manufacture, use, or sale of drugs" meant the entirety of the Food, Drug and Cosmetic Act, which regulates drugs but also covers medical devices and other products. As a result of the *Eli Lilly and Co. v. Medtronic* holding, 35 U.S.C. § 271(e)(1) extends to a range of subject matter, including pharmaceuticals, medical devices, food additives, color additives, and biological products. A generic firm may therefore use these products during the term of another's patent without fear of infringement liability, as long as the use is reasonably related to obtaining data for FDA approval.[105]

Other noteworthy judicial opinions have considered the nature of the activities that are exempted by the Hatch-Waxman Act privilege. The leading decision in *Intermedics, Inc. v. Ventritex, Inc.* interpreted the statutory exemption generously.[106] According to the *Intermedics* decision, "[w]here it would have been reasonable, objectively, for an accused infringer to believe that there was a decent prospect that the use in question would contribute (relatively directly) to the generation of information that was likely to be relevant in the processes by which the FDA would decide to approve the product,"[107] then the court should apply the 35 U.S.C. § 271(e)(1) infringement exemption. Following *Intermedics*, a number of other decisions have found that the accused infringer's activities fall within the statutory experimental use exemption.[108]

The Hatch-Waxman experimental use privilege is not without limit, however. A 2003 Federal Circuit decision, *Integra Lifesciences, I, Ltd. v. Merck*,[109] held that the 35 U.S.C. § 271(e)(1) experimental use exemption did not apply under the facts before the court. Here, Integra sued Merck for infringement of several patents relating to compounds thought to eliminate tumor growth and for treating a variety of other diseases. In turn, Merck asserted 35 U.S.C. § 271(e)(1) as a defense. The district court held that the Scripps-Merck activity did not fall under the 35 U.S.C. § 271(e)(1) exemption.[110]

The Federal Circuit affirmed the district court on appeal. The Federal Circuit observed that Merck's experiments did not supply information for submission to the FDA. Rather, Merck conducted these experiments in order to determine which of several compounds was the best drug candidate to subject to future testing.[111] The court then reasoned that such activities were not conducted "solely for purposes reasonably related to the development and submission of information under federal law," as the statutory exemption required. Rather, they were early-stage, exploratory experiments performed merely to identify promising pharmaceutical compounds.[112] The *Integra Lifesciences* decision makes clear that the 35 U.S.C. § 271(e)(1) exemption does not apply to all exploratory research, but is instead a "narrowly tailored" exemption intended to have a "*de minimis* impact on the patentee's right to exclude."[113]

Other Statutes

Although the Hatch-Waxman Act awards limited experimental use privileges with respect to patents on certain drugs, medical devices, and other products regulated by the FDA, the Patent Act of 1952 does not provide for a more broadly oriented experimental use privilege. However, Congress has enacted a number of specialized intellectual property statutes that expressly incorporate a more comprehensive experimental use doctrine. For example, the Plant Variety Protection Act allows the Department of Agriculture to issue plant variety certificates on sexually reproducible plants.[114] These certificates provide their owner with the exclusive right to "exclude others from selling the variety, or offering it for sale, or reproducing it, importing, or exporting it, or using it in producing (as distinguished from developing) a hybrid or different variety therefrom."[115] The statute does include a research exemption, however, stipulating that "[t]he use and reproduction of a protected variety for plant breeding or other bona fide research shall not constitute an infringement of the protection provided under this chapter."[116]

Another specialized intellectual property statute, the Semiconductor Chip Protection Act of 1984,[117] allows individuals to claim exclusive rights in "mask works" — circuitry designs used on a computer or other semiconductor chip.[118] It is an act of infringement either to reproduce these mask works, or to sell or import a semiconductor chip embodying a protected mask work.[119] However, this legislation expressly exempts those individuals who "reproduce the mask work solely for the purpose of teaching, analyzing, or evaluating the concepts or techniques embodied in the mask work or the circuitry, logic flow, or organization of components used in the mask work."[120]

A third example is provided by the Vessel Hull Design Protection Act.[121] This legislation, which allows designers of an original boat hull to register their designs with the federal government, incorporates an experimental use privilege similar to that of the Semiconductor Chip Protection Act. Under the Vessel Hull Design Protection Act, individuals who make, use, sell or import the protected design without the authorization of the registered design owner may be subject to infringement liability.[122] Yet the statute expressly exempts from infringement those uses "solely for the purpose of teaching, analyzing, or evaluating the appearance, concepts, or techniques embodied in the design, or the function of the useful article embodying the design."[123]

Legislative Issues and Alternatives

Should congressional interest continue in this area, a variety of options are available. If the current scope of the common law experimental use privilege is deemed to be appropriate, then no action need be taken. Alternatively, Congress could enact legislation confirming the narrowly cabined view of the experimental use privilege as set forth in *Madey v. Duke University* and predecessor cases.

If reform of the experimental use privilege is deemed prudent, however, another possibility is the introduction of some additional form of the experimental use privilege into the Patent Act of 1952. This infringement exemption could supplement or replace the narrow experimental use privilege introduced by the Hatch-Waxman Act. One option is to incorporate a generally applicable privilege of the sort contemplated by the proposed, but unenacted Patent Competitiveness and Technological Innovation Act of 1990. H.R. 5598 was introduced before the 101st Congress on September 12, 1990. Section 402 of that bill provided:

> It shall not be an act of infringement to make or use a patented invention solely for research or experimentation purposes unless the patented invention has a primary purpose of research or experimentation. If the patented invention has a primary purpose of research or experimentation, it shall not be an act of infringement to manufacture or use such invention to study, evaluate, or characterize such invention or to create a product outside the scope of the patent covering such invention.

H.R. 5598 was reported by the House Judiciary Committee on October 26, 1990. Other intellectual property statutes, including the Plant Variety Protection Act, Semiconductor Chip Protection Act, and the Vessel Hull Design Protection Act, also provide examples of the manner in which a generally applicable statutory experimental use privilege could be drafted.

A more limited statutory experimental use privilege presents another law reform option. Such a privilege might be limited to patented inventions that arise in particular technological fields, in the fashion of the proposed, but unenacted Genomic Research and Diagnostic Accessibility Act of 2002. H.R. 3967 would have exempted from patent infringement the use of genetic sequence information for purposes of research. In addition, this bill would have limited the remedies that the owner of a patent on genetic diagnostic testing could obtain during infringement litigation. Following its introduction in the 107th Congress on March 14, 2002, H.R. 3967 was referred to the Subcommittee on Courts, the Internet and Intellectual Property on May 6, 2002, but no further action was taken.

Another possibility would be to limit the experimental use privilege to patented research tools, in whatever technological field they might arise. Yet another option is to grant an experimental use privilege in favor of universities or non-profit research institutions, but retain the current law of experimental use with respect to for-profit enterprises.

Consideration of any sort of statutory reform with respect to experimental use privilege should take into account the Agreement on Trade-Related Aspects of Intellectual Property Rights.[124] The United States is a signatory to the so-called "TRIPS Agreement," which is a component of the international agreements that form the World Trade Organization (WTO). The TRIPS Agreement in part requires its signatories to grant patent owners the right to exclude others from making, using, offering for sale, selling, or importing a patented invention.[125] Further, under Article 27 of the TRIPS Agreement, such rights must be "enjoyable without discrimination as to the place of invention, the field of technology and whether products are imported or locally produced."

The TRIPS Agreement does allow member states to limit patent rights under certain circumstances, however. As stated in Article 30 of the TRIPS Agreement:

> Members may provide limited exceptions to the exclusive rights conferred by a patent, provided that such exceptions do not unreasonably conflict with a normal exploitation of the patent and do not unreasonably prejudice the legitimate interests of the patent owner, taking account of the legitimate interests of third parties.

Article 30 presumably allows its signatories to provide for a generally applicable experimental use privilege. Many WTO members, including Germany,[126] Japan[127] and the United Kingdom,[128] already incorporate such a privilege into their patent statutes. However, a more limited form of the experimental use privilege may raise concerns under the non-discrimination provision of Article 27. To the extent that the experimental use privilege is available for some sorts of inventions and not others, it may conflict with the Article 27 obligation not to discriminate as to the "field of technology" in which a patentable invention arises.[129]

Finally, the experimental use privilege need not be an all-or-nothing proposition. Another option is to grant researchers the ability to experiment with the patented inventions of others — provided they compensate the patent holder at a specified royalty rate. This regime would effectively amount to a "compulsory license" available to researchers.[130] Janice Mueller, a member of the faculty of the University of Pittsburgh Law School, posits that this approach would ensure "a royalty award of sufficient amount to maintain incentives for the development and patenting of new research tools, yet [alleviate] the access restrictions and up-front costs currently associated with acquisition and use of many proprietary research tools."[131]

In weighing this approach to the experimental use issue, it is also important to note that the TRIPS Agreement places some restrictions upon the ability of WTO members to grant compulsory licenses. Article 31, which is among the more detailed provisions in the TRIPS Agreement, in part requires that each application for a compulsory license be considered on its individual merits; that the proposed user must have made efforts to obtain a license from the patent owner; and that the legal validity of such a license be subject to review by the courts or other independent authority.[132] Deliberations over a "compulsory license" approach to the experimental use privilege may wish to account for these obligations.

CONCLUDING OBSERVATIONS

Whether the patent law's experimental use privilege should be retained as a narrowly confined doctrine of limited availability, or expanded to encompass additional experimental activities, technologies, and researchers, continues to be the subject of active discussion in the scientific and legal communities. A limited experimental use privilege may best encourage technological advancement by rewarding successful researchers with patent rights that are not easily circumvented. However, some commentators believe that the circumscribed nature of the experimental use privilege may in fact restrict researcher access to state-of-the-art technologies and thus discourage further technological development. Although the courts have relied upon existing law in order to reach their decisions in particular cases, whether a narrow experimental use privilege most appropriately serves the contemporary scientific research community remains open to policy debate

REFERENCES

[1] See CRS Report 95-50, The Federal Role in Technology Development, by Wendy H. Schacht.

[2] See CRS Issue Brief IB10129, Federal Research and Development Funding: FY2005, by Michael E. Davey, coordinator.

[3] See CRS Report RL32076, The Bayh-Dole Act: Selected Issues in Patent Policy and the Commercialization of Technology, by Wendy H. Schacht; and CRS Report RL32324, Federal R&D, Drug Discovery, and Pricing: Insights from the NIH-University-Industry Relationship, by Wendy H. Schacht.

[4] Janice M. Mueller, "No 'Dilettante Affair': Rethinking the Experimental Use Exception to Patent Infringement for Biomedical Research Tools," 76 Washington Law Review (2001), 1.

[5] Michelle Cai, "Madey v. Duke University: Shattering the Myth of Universities' Experimental Use Defense," 19 Berkeley Technology Law Journal (2004), 175.

[6] Douglas v. United States, 181 U.S.P.Q. (BNA) 170, 176 (Ct. Cl. Tr. Div. 1974), aff'd on other grounds, 184 U.S.P.Q. (BNA) 613 (Ct. Cl. 1975).

[7] Whittemore v. Cutter, 29 F. Cas. 1120 (C.C. Mass. 1813).

[8] See CRS Report RL30756, Patent Law and Its Application to the Pharmaceutical Industry: An Examination of the Drug Price Competition and Patent Term Restoration Act of 1984 ("The Hatch-Waxman Act"), by Wendy H. Schacht and John R. Thomas.

[9] See Katherine J. Strandburg, "What Does the Public Get? Experimental Use and the Patent Bargain," 2004 Wisconsin Law Review 81.

[10] 307 F.3d 1351 (Fed. Cir. 2002).

[11] See Jennifer Miller, "Sealing the Coffin on Experimental Use," 2003 Duke Law & Technology Review (May 7, 2003), 12.

[12] Lawrence M. Sung & Claire M. Maisano, "Piercing the Academic Veil: Disaffecting the Common Law Exception to Patent Infringement Liability and the Future of a Bona Fide Research Use Exemption After Madey v. Duke University," 9 Journal of Health Care Law and Policy (2003), 256.

[13] Jordan P. Karp, "Experimental Use as Patent Infringement: The Impropriety of a Broad Exception," 100 Yale Law Journal 2169 (1991).
[14] Rebecca S. Eisenberg, "Patents and the Progress of Science: Exclusive Rights and Experimental Use," 56 University of Chicago Law Review 1017 (1989).
[15] See, e.g., David C. Hoffman, "A Modest Proposal: Toward Improved Access to Biotechnology Research Tools By Implementing a Broad Experimental Use Exception," 89 Cornell Law Review (2004), 993.
[16] Roger E. Schechter & John R. Thomas, Principles of Patent Law § 1.3.1 (Thomson West 2004).
[17] See, e.g., Grant v. Raymond, 31 U.S. 218, 247 (1832).
[18] R. Polk Wagner, "Information Wants to Be Free: Intellectual Property and the Mythology of Control," 103 Columbia Law Review (2003), 995.
[19] F. Scott Kieff, "Property Rights and Property Rules for Commercializing Inventions," 85 Minnesota Law Review (2000), 697.
[20] See Robert P. Merges, "Intellectual Property and the Costs of Commercial Exchange: A Review Essay," 93 Michigan Law Review (1995), 1570.
[21] See Frederic M. Scherer & David Ross, Industrial Market Structure and Economic Performance (Rand McNally & Co., 3d ed. 1990).
[22] See, e.g., Pamela Samuelson, Benson Revisited: The Case Against Patent Protection for Algorithms and Other Computer Program — Related Inventions, 39 Emory Law Journal (1990), 1025.
[23] CRS Report RL31951, Innovation, Intellectual Property, and Industry Standards, by John R. Thomas.
[24] Pub. L. No. 82-593, 66 Stat. 792 (codified at Title 35 United States Code).
[25] 35 U.S.C. § 111 (2000).
[26] 35 U.S.C. § 131 (2000).
[27] John R. Thomas, "On Preparatory Texts and Proprietary Technologies: The Place of Prosecution Histories in Patent Claim Interpretation," 47 UCLA Law Review (1999), 183.
[28] 35 U.S.C. § 112 (2000).
[29] Ibid.
[30] 35 U.S.C. § 101 (2000).
[31] 35 U.S.C. § 102 (2000).
[32] 35 U.S.C. § 103 (2000).
[33] 35 U.S.C. § 271(a) (2000).
[34] 35 U.S.C. § 154(a)(2) (2000).
[35] 35 U.S.C. § 281 (2000).
[36] 28 U.S.C. § 1295(a)(1) (2000).
[37] 28 U.S.C. §1254(1) (2000).
[38] 35 U.S.C. § 271(a) (2000).
[39] 35 U.S.C. § 282 (2000).
[40] 29 F.Cas. 1120 (C.C.D.Mass.1813) (No. 17,600).
[41] 29 F. Cas. at 1121.
[42] 21 F. Cas. 554 (C.C.D. Mass 1813) (No. 12, 391).
[43] 19 F. Cas. 1048, 1049 (C.C.S.D.N.Y. 1861).

[44] James Boyle, "Foreward: The Opposite of Property," Law & Contemporary Problems (Winter/Spring 2003), 27.
[45] Kevin Sandstrom, "How Much Do We Value Research and Development: Broadening the Experimental Use Exemption to Patent Infringement in Light of Integra Lifesciences I Ltd. v. Merck KGaA, 30 William Mitchell Law Review (2004), 1067.
[46] Rebecca S. Eisenberg, "Proprietary Rights and the Norms of Science in Biotechnology Research," 97 Yale L. J. 177, 220 (1987).
[47] See, e.g., Radio Corp. of America v. Andrea, 15 F.Supp. 685, 30 USPQ 194 (E.D. N.Y. 1936), modified, 90 F.2d 612, 34 USPQ 312 (2d Cir. 1937).
[48] See, e.g., Cimotti Unharing Co. Derboklow, 87 F. 997, 999-1000 (C.C.E.D.N.Y. 1898).
[49] 547 F.2d 1106 (Ct. Cl. 1976).
[50] 547 F.2d at 1125-26.
[51] See also, e.g., Embrex Inc. v. Service Engineering Corp., 216 F.3d 1343 (Fed. Cir. 2000).
[52] 307 F.3d 1351 (Fed. Cir. 2002).
[53] 307 F.3d at 1352-53.
[54] 266 F. Supp. 2d 420 (M.D.N.C. 2001).
[55] 307 F.3d at 1361.
[56] 307 F.3d at 1362.
[57] Ibid.
[58] 307 F.3d at 1362-63.
[59] See Tom Saunders, "Renting Space on the Shoulders of Giants: Madey and the Future of the Experimental Use Doctrine," 113 Yale Law Journal (2003), 261 (concluding that Madey "reformulated the experimental use doctrine and cast considerable doubt on its continued viability as a defense in patent infringement cases involving universities").
[60] Sung & Maisano, supra note 10.
[61] Cai, supra note 5.
[62] See Michael R. Taylor & Jerry Cayford, "American Patent Policy, Biotechnology, and African Agriculture: The Case for Policy Change," 17 Harvard Journal of Law & Technology (2004), 321 ("In a recent decision, the U.S. Court of Appeals for the Federal Circuit narrowed the [common law experimental use] exemption to the point of eliminating it for practical purposes.").
[63] Paul Devinksy & Mark G. Davis, "2003 Patent Law Decisions of the Federal Circuit," 53 American University Law Review (2004), 773, 883.
[64] Cai, supra note 5.
[65] Cathryn Campbell & R.V. Lupo, "Exemption to Patent Infringement Under 35 U.S.C. Section 271(e)(1): Safe Harbor or Storm A-Brewing?," 5 Sedona Conference Journal (2004), 29.
[66] Karp, supra note 13.
[67] Ibid.
[68] Cai, supra note 5.
[69] Stephen G. Kunin & Linda S. Therkorn, "Workship on Future Public Policy and Ethical Issues Facing the Biotechnology Industry," 86 Journal of the Patent and Trademark Office Society (2004), 503.
[70] Act of Dec. 12, 1980, Pub. L. No. 96-517, 94 Stat. 3015.
[71] See Schacht, supra note 3.

[72] See Traci Dreher Quigley, "Commercialization of the State University: Why the Intellectual Property Restoration Act of 2003 Is Necessary," 152 University of Pennsylvania Law Review (2004), 2001.
[73] See Eisenberg, supra note 14.
[74] See Maureen A. O'Rourke, "Toward a Doctrine of Fair Use in Patent Law," 100 Columbia Law Review (2000), 1177.
[75] John P. Walsh et al., "Working Through the Patent Problem," 299 Science 1021, 1021 (2003) ("[Research universities] have largely ignored the growing number of patents covering technology that their scientists use without license and without apology.").
[76] Miller, supra note 11.
[77] Natalie M. Derzko, "In Search of a Compromised Solution to the Problem Arising From Patenting Biomedical Research Tools," 20 Santa Clara Computer & High Technology Law Journal (2004), 347.
[78] Michael A. Heller & Rebecca S. Eisenberg, Can Patents Deter Innovation? The Anticommons in Biomedical Research, 280 Science 698 (1998).
[79] Ibid.
[80] Strandburg, supra note 9.
[81] F. Scott Kieff, "Facilitating Scientific Research: Intellectual Property Rights and the Norms of Science — A Response to Rai and Eisenberg," 95 Northwestern Law Review (2001), 691.
[82] J. Walsh, A. Arora and W.M. Cohen, "The Patenting and Licensing of Research Tools and Biomedical Innovation," Mimeo, National Academy of Sciences, October, 2001.
[83] Heather Hamme Ramirez, "Defending the Privatization of Research Tools: An Examination of the "Tragedy of the Anticommons" in Biotechnology Research and Development," 53 Emory Law Journal (2004), 362.
[84] Department of Health and Human Services, National Institutes of Health, "Principles and Guidelines for Recipients of NIH Research Grants and Contracts on Obtaining and Disseminating Biomedical Research Resources," 64 Fed. Reg. 72,090 (Dec. 23, 1999).
[85] Id. at 72,090.
[86] See, e.g., Kimberly A. Moore, "Are District Court Judges Equipped to Resolve Patent Cases?," 15 Harvard Journal of Law and Technology (2001), 1.
[87] Andrew J. Caruso, "The Experimental Use Exception: An Experimentalist's View," 14 Albany Journal of Law, Science & Technology (2003), 217.
[88] CRS Report RL32400, Patents and Drug Importation, by John R. Thomas.
[89] Edward T. Lentz, "Pharmaceutical and Biotechnology Research After Integra and Madey," 23 Biotechnology Law Report (2004), 265.
[90] 35 U.S.C. § 271(a) (2000).
[91] Pub. L. No. 98 — 417, Title II, 98 Stat. 1585 (Sept. 28, 1984).
[92] CRS Report RL30989, The U.S. Drug Approval Process: A Primer, by Blanchard Randall IV.
[93] See Shashank Upadhye, "Understanding Patent Infringement Under 35 U.S.C. § 271(e)(1): The Collisions Between Patent, Medical Device, and Drug Laws," 17 Santa Clara Computer & High Technology Law Journal (2000), 1.
[94] See Janet A. Gongola, "Prescriptions for Change: The Hatch-Waxman Act and New Legislation to Increase the Availability of Generic Drugs to Consumers," 36 Indiana Law Review (2003), 787.

[95] 35 U.S.C. § 271(a) (2000).
[96] Alfred B. Engelberg, "Special Patent Provisions for Pharmaceuticals: Have They Outlived Their Usefulness?," 39 IDEA: Journal of Law and Technology (1999), 389.
[97] 733 F.2d 858 (Fed. Cir. 1984).
[98] 733 F.2d at 860.
[99] 733 F.2d at 863.
[100] See CRS Report RL30756, Patent Law and Its Application to the Pharmaceutical Industry: An Examination of the Drug Price Competition and Patent Term Restoration Act of 1984 ("Hatch-Waxman Act"), by Wendy H. Schacht and John R. Thomas.
[101] 21 U.S.C. § 355(j) (2000).
[102] Pub. L. No. 100-670, 102 Stat. 3971.
[103] 496 U.S. 661 (1990).
[104] 496 U.S. at 665-66.
[105] See Patricia Nussle, "Eli Lilly & Co. v. Medtronic Inc.: A Case of Statutory Interpretation," 54 Ohio State Law Journal (1992), 645.
[106] 775 F.Supp. 1269 (N.D. Cal. 1991), aff'd, 991 F.2d 808 (Fed.Cir.1993).
[107] 775 F. Supp. at 1280.
[108] See, e.g., Abtox, Inc. v. Exitron Corp., 122 F.3d 1019 (Fed. Cir. 1997); Glaxo, Inc. v. Novopharm, Ltd.,110 F.3d 1562 (Fed. Cir.1997); Amgen, Inc. v. Hoechst Marion Roussel, Inc., 3 F.Supp.2d 104 (D. Mass.1998).
[109] 331 F.3d 860 (Fed. Cir. 2003).
[110] Integra Lifesciences I Ltd. v. Merck KGaA, 50 USPQ2d 1846 (S.D. Cal. 1999).
[111] 331 F.3d at 863.
[112] 331 F.3d at 866.
[113] 331 F.3d at 867.
[114] Pub. L. No. 91-577, 84 Stat. 1542 (1970) (codified at 7 U.S.C. §§ 2321-2583). See CRS Report RL31568, Plants, Patents and Seed Innovation in the Agricultural Industry, by John R. Thomas.
[115] 7 U.S.C. § 2483(a).
[116] 7 U.S.C. § 2544 (2000).
[117] Pub. L. No. 98-620, 98 Stat. 3347.
[118] 17 U.S.C. § 901 (2000).
[119] 17 U.S.C. § 905 (2000).
[120] 17 U.S.C. § 906(a)(1) (2000). See Lee Hsu, "Reverse Engineering Under the Semiconductor Chip Protection Act: Complications for Standard of Infringement," 5 Albany Law Journal of Science and Technology (1996), 249.
[121] Pub. L. No.105-304 , 112 Stat. 2905.
[122] 17 U.S.C. § 1308 (2000).
[123] 17 U.S.C. § 1309(g) (2000). The Digital Millennium Copyright Act provides another example of a statutory experimental use privilege. See 17 U.S.C. § 1201 (2000).
[124] Agreement on Trade-Related Aspects of Intellectual Property Rights, Apr. 15, 1994, Annex 1C, 33 I.L.M. 1197 (1994) ("TRIPS Agreement").
[125] TRIPS Agreement, Article 27(1).
[126] German Patent Act, Article 11, § 2. See Klinische Versuche (Clinical Trials) II (Case X ZR 68/94), Federal Supreme Court of Germany, BGHZ 135, 217, [1998] R.P.C. 423.
[127] Japanese Patent Act § 69(1), available at [http://www.jpo.go.jp/shoukaie/patent.htm].

[128] United Kingdom Patent Act, 1977, ch. 37, § 60(5)(b) (Eng.), reprinted in U.K. Patent Office, Manual of Patent Practice (1999), available at [http://www.patent.gov.uk/-patent/reference/mpp/s60_71.pdf] (providing a defense to infringement for actions "done for experimental purposes relating to the subject-matter of the invention").

[129] Notably, a Canadian law modeled after the Hatch-Waxman Act was challenged before a WTO tribunal on precisely this ground. The WTO dispute resolution panel upheld the Canadian experimental use privilege based upon the particular circumstances created by the marketing approval and patenting procedures. See Canada — Patent Protection of Pharmaceutical Products, WTO Panel Report, WT/DS114/R (Mar. 17, 2000). Some commentators believe that Article 30 was "certainly interpreted narrowly by the WTO Panel." See Sol Picciotto, "Private Rights vs. Public Interests in the TRIPS Agreement," 97 American Society of International Law Proceedings (April 2-5 2003), 167. Whether other narrowly tailored experimental use privileges would survive scrutiny under the TRIPS Agreement remains uncertain.

[130] See CRS Report RL31132, Multinational Patent Acquisition and Enforcement: Public Policy Challenges and Opportunities for Innovative Firms, by John R. Thomas.

[131] See Mueller, supra note 4.

[132] Schechter & Thomas, supra note 16 , at § 12.6.

In: New Intellectual Property Issues
Editor: John V. Luong, pp. 57-73
ISBN: 978-1-60021-211-6
© 2007 Nova Science Publishers, Inc.

Chapter 3

INTELLECTUAL PROPERTY, COMPUTER SOFTWARE AND THE OPEN SOURCE MOVEMENT*

John R. Thomas
Scholar Resources, Science, and Industry Division

SUMMARY

The term "open source" refers to a computer program that is distributed along with a license, or contract, that requires users of the program to comply with specified conditions. Among these stipulations are that the source code be distributed along with the software, and that others be allowed to modify the source code as they desire. In contrast, the source code of "closed source" software is proprietary, not publicly distributed and subject to alteration only by the software manufacturer.

Some concerns have arisen concerning the relationship between open source software and intellectual property rights, including copyrights, patents and trade secrets. Although a particular computer program may be designated as open source, it remains possible that an owner of intellectual property may enforce its rights against open source software developers and users. Some commentators have also expressed concern that open source licenses may overreach, converting proprietary programs into open source software even if only a portion of that program was derived from an open source original. Others have suggested that open source licenses may not be legally enforceable, which would allow users to obtain and assert intellectual property rights pertaining to software that was initially distributed as open source.

Striking a balance between promoting innovation, on one hand, and accommodating the demands of software developers and users, on the other, forms an important component of contemporary software policy. The possibility of intellectual property rights, and their attendant license fees and royalties, may provide a significant incentive for firms to innovate and to distribute software. However, some proponents of open source software believe that these incentives are unnecessary, and further hope to maintain a non-proprietary environment of software distribution and development.

Should Congress have an interest in this area, several options present themselves. No action need be taken if the current relationship between open source software and intellectual property is deemed satisfactory, particularly as software publishers become increasingly aware of intellectual property and as judicial precedents may make the legal situation clearer. Congress might also assist individuals in identifying intellectual property that pertains to

* Excerpted from CRS Report RL32268, Updated March 11, 2004

software that has been identified as open source; speak to the enforceability of open source licenses; and, as a possible more far-reaching legal reform, allow proprietary software publishers a grace period for removing portions of program code that derived from an open source original.

Perhaps the most dramatic development in the contemporary computer industry is the "open source" software movement. The term "open source" refers to a computer program whose "source code" is made available to the public for modification or improvement as individual users desire.[1] In contrast, the source code of "closed source" software is proprietary, not publicly distributed and subject to alteration only by the software manufacturer.

The rise of open source software has generated considerable discussion in recent years. Proponents of open source software contend that the open source system preserves the freedom of computer users and provides a superior development methodology as compared to the usual proprietary model.[2] Other commentators have expressed concerns regarding the security of open source software and whether it will operate in a compatible fashion with other programs.[3]

Intellectual property rights, including copyrights, patents and trade secrets, present another possible set of concerns with respect to open source software. Although a particular computer program may be designated as open source, it remains possible that an owner of intellectual property may enforce its rights against open source software developers and users.[4] In addition, open source software is ordinarily accompanied by a license that requires users to maintain the program as open source.[5] Some commentators have expressed concern that these licenses may overreach, converting proprietary programs into open source software even if only a portion of that program was derived from an open source original.[6] Others have suggested that open source licenses may not be legally enforceable, which would allow users to obtain and assert intellectual property rights pertaining to software that was initially distributed as open source.[7]

This report considers the impact of intellectual property rights upon open source software. This report commences with an introduction to the open source movement in the software industry. Next, it briefly reviews the intellectual property laws, including copyrights, patents and trade secrets. After identifying issues of interface between open source software and the intellectual property laws, this report concludes with a discussion of possible legislative issues and approaches.

Introduction to the Open Source Movement

Fundamental Concepts

An understanding of the fundamentals of modern computer software technology will assist in an understanding of the open source movement. Software programmers typically write software programs using a high level computer language such as Basic, C++, or Java. By using the words, symbols and numbers that make up these high-level computer languages, the programmer tells the computer what to do. For instance, the command "ADD (X, Y)" instructs the computer to add the value of the variable X to the variable Y. A computer program written in this high level language is said to be in "source code" form.[8]

Computers are incapable of reading the high-level instructions of source code, however. Rather, the computer responds to binary inputs — either a "0" or a "1" —that correspond to an open or closed electrical switch. The number "01001101" might tell a computer to add two numbers and save the result, for example. Thus, after writing the program in source code, the programmer ordinarily uses a compiler program to translate or "compile" the source code into the corresponding 1s and 0s that the computer can read. Source code that has been compiled into this series of 1s and 0s is called "object code."[9]

Today, most computer programs are distributed in object code form. This distribution system is often referred to as the "proprietary software" model. When a consumer purchases a copy of, for example, WordPerfect software, that program comes on a CD-ROM into which the object code 1s and 0s have been encoded. One of the reasons that software developers distribute only the object code, and not the source code, is convenience. Software in object code format is ready for the computer to use.[10]

This technique also protects the source code from disclosure. Because skilled programmers can easily read source code, a competitor could review this text in order to find out how a program works. As a result, if software were distributed in source code form, rival firms could readily take and reuse parts of the program in competing products. As this report discusses below,[11] this appropriation of the original programmer's work may violate intellectual property laws. However, such violations may be difficult both to discover and stop. Distributing software in object code form is a more cost-efficient and effective means of preventing this infringement.[12]

Some computer users have expressed dissatisfaction with the proprietary software model. End users of proprietary software must ordinarily rely upon the software publisher to fix mistakes in the code and develop additional features. As free software advocate Richard Stallman has explained, the proprietary software model "keeps users helpless and divided: the inner workings are secret."[13] Stallman further opines that computer users "should be free to modify programs to fit their needs, and free to share software, because helping other people is the basis of society."[14]

A loosely organized open source software community has resulted from this reaction to proprietary software development.[15] In contrast to proprietary software, the source code of "open source" software is available to the public.[16] Computer users are therefore able to examine the human-readable instructions that the programmer wrote to create the software. A number of well-known programs are "open source," including a widely used Internet server program, Apache; a popular Internet programming language, Perl; the program that routes more than 80 percent of all Internet email messages worldwide, Sendmail; the program that is the basis for the domain name system, Bind; and the fastest growing operating system in the world, Linux.[17] Due to the open source status of these programs, members of the public may consult their source code as desired.

Open Source Software Licenses

It is important to note that the term "open source" implies more than merely distribution of source code along with the object code. In such circumstances, the publisher could continue to assert its intellectual property rights in the software, thereby limiting the ability of others to modify the program or redistribute it. Even though disclosed to the public, the source code would remain under the control of the publisher.[18]

As a result, publishers of open source software ordinarily do more than simply provide copies of both the source code and the object code when they distribute computer programs to the public. In addition, they establish the terms of use of the software by means of a license. A license is a contract through which the publisher allows recipients to use and modify the software, subject to certain conditions specified in the license. For example, the license might require that anyone who redistributes the software also make the source code of that software publicly available. Contracts that provide users with a sufficient set of privileges to access and modify the software's source code are deemed to be "open source licenses."

The practice of preserving the rights of software users through a set of license provisions is sometimes called "copylefting." This term is a play on words on the term "copyright."[19] Under this system, the copyright holder licenses the recipient of a copy of the software. The license permits the redistribution of further copies of the software — including software containing modifications — under the condition that those copies are subject to the same license. This legal framework ensures that derivatives of the licensed work remain open. If the licensee fails to distribute derivative works under the same license, then he may face legal consequences. In particular, the licensor could terminate the license, leaving the licensee without permission to copy, distribute, or modify the software. [20]

Although no official definition specifies which software licenses qualify as open source licenses, an organization called the "Open Source Initiative," or OSI, has promulgated a widely followed set of standards.[21] OSI describes itself as a "nonprofit corporation dedicated to managing and promoting the Open Source Definition"[22] through a certification program that it administers.[23] To satisfy the Open Source Definition, the license must satisfy certain conditions, including:

1. The publisher must provide both object and source code.
2. The publisher must allow modification and redistribution of the code (with or without modifications by the licensee).
3. The publisher must not limit distribution to certain fields of endeavor or products, or limit its use with other free software. [24]

A number of different groups have promulgated a variety of open source licenses that OSI has certified as compliant with the Open Source Definition. Among these is the General Public License, or GPL.[25] As compared to the Open Source Definition, the GPL imposes additional restrictions upon software publishers. As a result, a license may fulfill the conditions of the Open Source Definition but not qualify as a GPL. According to its sponsor, the Free Software Foundation, the GPL guarantees computer users the following "four freedoms":

> The freedom to run the program, for any purpose.
> The freedom to study how the program works, and adapt it to your needs.
> The freedom to redistribute copies so you can help your neighbor.
> The freedom to improve the program, and release your improvements to the public, so that the whole community benefits.[26]

It is important to note that, in this context, the term "freedom" does not mean that the software has to be sold at no charge. Rather, it refers to permissible user activities, and in particular the principle that software should be openly available in all of its current and future forms to all those desiring to learn or benefit from it.[27]

The chief distinction between the Open Source Definition and the GPL is that the Open Source Definition effectively allows users to appropriate privately any modifications that they make.[28] For example, under the Open Source Definition, a user could refuse to disclose publicly the source code of any programs that include the user's modifications. The licensee could also claim intellectual property rights in any modifications they introduced into the software.[29] Of course, in such cases, the licensee's software would not qualify as open source. In contrast, the GPL requires that the source code be kept open to the public, even if the recipient of the software made changes. Also, under the GPL, the licensee cannot restrict the ability of others to build upon any modifications that the licensee made.[30]

For example, a computer programmer may use software under a license that minimally complies with the Open Source Definition. Suppose further that the programmer modifies the software, and then licenses the object code of the modified version with the additional restriction that no licensee could copy the modified version's object code. This practice is acceptable under the Open Source Definition. This practice would not comply with the Free Software Foundation's GPL, however.

Introduction to Intellectual Property

As we have seen, both proprietary and open source software publishers rely upon the intellectual property laws to sustain their chosen development model. More specifically, the term "intellectual property laws" identifies a number of doctrines, including copyrights, patents and trade secrets, that provide innovators with proprietary interests in their intangible creations.[31] Copyright provides authors with exclusive rights in their writings, visual works and other works of authorship; patents relate to products, processes and other useful inventions; while trade secret law concerns secret information that is of commercial value.[32] After briefly identifying the principal policy justifications for the copyright, patent and trade secret law, this report summarizes the basic workings of these legal disciplines.

Intellectual Property Policy

By providing individuals with exclusive rights to their inventive products and processes, the intellectual property laws allows authors and inventors to secure the economic benefits of their discoveries. Absent the a system of intellectual property, competitors might readily be able to appropriate the benefits of an innovator's research and development efforts. Aware of these potential "free riders," firms might devote few, if any resources towards innovation. Copyrights, patents and trade secrets solve this market failure problem by providing economic incentives for individuals and institutions to engage in research and development.[33]

Intellectual property rights may also facilitate market mechanisms by creating discrete, well-defined property interests that are readily subject to commercial exchanges.[34] Absent patent rights, for example, an inventor may have no tangible asset to sell or license. In addition, an inventor might otherwise be unable to police the conduct of a contracting party. Any technology or know-how that has been disclosed to a prospective buyer might be appropriated without compensation to the inventor. The availability of patent protection decreases the ability of contracting parties to engage in opportunistic behavior. By lowering such transaction costs, the patent system may make technology-based transactions more feasible.[35]

Some commentators have identified natural rights as an additional rationale for the intellectual property laws, and in particular the copyright law. Under this view, individuals enjoy a natural right of property in the works they have created through their own labors.[36] Intellectual property law may also encourage marketplace behavior that is both ethical and efficient. As a matter of fairness and justice, this view holds, one individual should not be allowed to profit from another's labor and goodwill.[37] In economic terms, intellectual property laws may also stimulate socially desirable activities that create wealth, including innovation, and discourage costly acts that merely redistribute existing wealth, such as industrial espionage.[38]

The intellectual property system has long been subject to criticism, however. Some observers believe that intellectual property rights encourage industry concentration and create barriers to market entry.[39] Others have suggested that intellectual property rights can actually slow the pace of innovation in some circumstances, as competitors are discouraged from pursuing certain research pathways due to existing proprietary interests.[40] And in an era where digital information can be immediately disseminated across the globe, the notion that an innovation can be an object of possession has been challenged.[41]

When analyzing these contending views, it is important to note the lack of rigorous analytical methods available for determining the effect of the intellectual property rights upon the U.S. economy as a whole. The relationship between innovation and intellectual property rights remains poorly understood. Concerned observers simply do not know what market impacts would result from changing the patent term from its current 20-year period, for example.[42] Consequently, current economic and policy tools do not allow us to calibrate the intellectual property system precisely in order to produce an optimal level of investment in innovation.

Copyrights

Under the federal Copyright Act of 1976, copyright may extend to any original work of authorship, including computer software programs.[43] Copyright protection arises automatically, without the need of the proprietor to take any formal steps. However, authors who register their works with the U.S. Copyright Office,[44] and then place a notice of copyright on copies of their works,[45] are afforded certain advantages when enforcing their copyrights.

Copyright confers a number of exclusive rights to the author or, in some circumstances, to the employer of the author.[46] Most notable is that the copyright proprietor has the exclusive right to make copies of the protected work and to distribute it to the public.[47] Each copyright ordinarily enjoys a term of the life of the author plus 70 years.[48]

As with the other intellectual property rights, copyrights do not enforce themselves. Copyright proprietors bear responsibility for monitoring its competitors to determine whether they are using the protected work or not. Copyright owners who wish to compel others to observe their intellectual property rights must usually commence litigation in the courts.

As with patents and trade secrets, copyrights may be sold or licensed to others.[49] An intellectual property license is a grant of permission to use the owner's proprietary interest. It does not convey ownership rights in the intellectual property, which remain with the licensor. A license is necessary if, without one, the licensee would be liable for infringement if it made use of the subject matter that is protected by an intellectual property right.

Patents

In contrast to copyright, patent rights do not arise automatically. Under the Patent Act of 1952, inventors must prepare and submit applications to the U.S. Patent and Trademark Office ("USPTO") if they wish to obtain patent protection.[50] USPTO officials known as examiners then assess whether the application merits the award of a patent.[51]

In deciding whether to approve a patent application, a USPTO examiner will consider whether the submitted application fully discloses and distinctly claims the invention.[52] In addition, the application must disclose the "best mode," or preferred way, that the applicant knows to practice the invention.[53] The examiner will also determine whether the invention itself fulfills certain substantive standards set by the patent statute. To be patentable, an invention must be useful, novel and nonobvious. The requirement of usefulness, or utility, is satisfied if the invention is operable and provides a tangible benefit.[54] To be judged novel, the invention must not be fully anticipated by a prior patent, publication or other knowledge within the public domain.[55] A nonobvious invention must not have been readily within the ordinary skills of a competent artisan at the time the invention was made.[56]

If the USPTO allows the patent to be issued, the patent proprietor obtains the right to exclude others from making, using, selling, offering to sell or importing into the United States the patented invention.[57] The maximum term of patent protection is ordinarily set at 20 years from the date the application is filed.[58] The patent applicant gains no enforceable rights until such time as the application is approved for issuance as a granted patent, however. Once the patent expires, others may employ the patented invention without compensation to the patentee.

Trade Secrets

The common law of trade secrets protects valuable, commercially useful information that is the subject of reasonable measures to preserve its secrecy.[59] Maintaining trade secret status ordinarily requires the trade secret proprietor to restrict access to the confidential information, require employees to promise not to disclose the information, or take other precautions that are prudent under the circumstances. Computer software may be the subject of trade secret protection so long as these requirements are met.

Unlike a copyright or patent, a trade secret is not limited in duration to a fixed number of years. Instead, a trade secret endures so long as the protected information remains commercially valuable and is maintained as a secret.[60] If the trade secret holder ceases to take measures to maintain the confidentiality of the protected information, or if the information becomes widely available to the public, then the trade secret is extinguished.

A trade secret is misappropriated when it has been obtained through the abuse of a confidential relationship or improper means of acquisition.[61] Typical cases of trade secret misappropriation involve trespass, bribery and theft, as well as breaches of a duty of confidence committed by former employees or commercial partners. Unlike the Patent Act, trade secret law does not provide a cause of action against an individual who independently developed or reverse engineered the subject matter of the trade secret.[62]

Potential Conflicts Between Open Source Software and Intellectual Property

Conflicts potentially arise between open source standards and intellectual property rights.[63] Some observers have expressed concerns that if open source software is incorporated into an otherwise proprietary program, then the terms of the open source license will apply to the entire program and defeat intellectual property rights that would otherwise exist. It is also possible that a party not bound by the terms of an open source license may raise claims of intellectual property infringement based upon the use of software that others believe to be open source. Finally, questions have arisen regarding the validity and enforceability of open source licenses. This report next reviews these issues.

The Alleged "Viral" Nature of Open Source Software

Certain open source licenses have sometimes been described as "viral" in character.[64] Some individual open source licenses require that the terms of that license apply automatically to each copy of the software, as well as to any modified versions.[65] Some observers have expressed concerns regarding situations where a programmer, perhaps unknowingly, incorporates some open source program code into a larger software package. In these circumstances, the open source portion of the software could "contaminate" the entire program. As a result, even though the publisher intended that the program be proprietary, it may be instead be distributed as the open source license stipulates.[66] In this way, the open source software component would trump any intellectual property rights that the publisher hoped to claim.

Whether a particular open source software is potentially "viral" or not depends upon the individual terms of the accompanying open source license. Some commentators have stated that the Free Software Foundation's GPL is one example of a potentially "viral" license.[67] Under the GPL, anyone who uses or modifies the software must, upon further distributing that software or a modification of that software, make the source code fully available to the public, free of any proprietary interest.[68] This limitation prevents the software written and distributed under the GPL from being subject to intellectual property rights. As a result, the GPL maintains the "open" nature of the open source code by allowing users to modify and redistribute the software, but requiring that such modifications be made available to anyone under the terms of the GPL.

Notably, under certain conditions the GPL allows users to incorporate open source-derived computer programs into proprietary software packages without subjecting the entire package to the GPL.[69] According to the Free Software Foundation, in order to avoid the imposition of the GPL upon the entire program, software publishers "must make sure that the free and non-free programs communicate at arms length, that they are not combined in a way that would make them effectively a single program."[70]

The Free Software Foundation offers as an example the combination of an editor program and a shell program.[71] To expand upon this illustration, suppose that an editor program executes textual commands typed by the user. A shell, or user interface program, might provide a graphical, menu-driven interface so that the user doesn't have to memorize the text commands. The editor could be a proprietary program. The shell might have been developed for individuals who did not wish to learn the text commands, and distributed under the GPL. As the editor can work independently of the shell in this example, the proprietary nature of the editor would be maintained even though the shell was subject to the GPL.[72]

The functional separation of proprietary and open source software provides one avenue for avoiding the broad application of an open source license to an entire software product. Some observers believe, however, that conformity with this exception imposes substantial compliance burdens. For example, attorneys James A. Harvey and Todd S. McClelland state:

> It is therefore a good practice to advise clients with respect to procedures designed to segregate software that can be licensed as proprietary from that which must be distributed under an open source license. Developing and maintaining procedures to implement these administrative tasks can be very difficult in complex development environments.[73]

Views differ on the supposed "viral" nature of some open source licenses. Representatives of proprietary software firms have expressed concerns that "open source is an intellectual-property destroyer,"[74] and have reportedly referred to open source software as a "cancer" and "un-American."[75] Others believe that, in order for open source software to remain open to the public, all programs derived from an open source original should be treated as open source as well.[76] And, as noted above, some open source licenses take an intermediate position.[77] Although these licenses require that the original software that is distributed with the license remain open source, they allow modifications and upgrades to that code to be "taken private" and be treated as proprietary software.

Third Party Infringement Claims

Some open source licenses, such as the General Public License (GPL), effectively prevent individuals from asserting intellectual property rights in open source software.[78] This restriction only applies to individuals who have consented to these licenses, however.[79] Parties not subject to that license are therefore not necessarily prevented from enforcing their intellectual property rights against individuals who use the software. As a result, even though one individual has distributed software that it has designated as open source, another entity may possibly assert that the software infringes an intellectual property right. Such assertions lead to potential conflicts between the intellectual property rights owner and individuals who believe that the software is open to the public.

For example, suppose that a computer scientist, Alpha, invents a new method of sorting data useful for computer programs. Alpha then files a patent application at the USPTO claiming the method. Later, a programmer named Beta independently writes a software program that uses the same data sorting method that Alpha had claimed in his pending application. Unaware of Alpha's patent application, Beta distributes his own software to the public under the GPL. If the USPTO approves of Alpha's application, Alpha could assert claims of patent infringement against anyone using Beta's software. These users would be subject to legal liability, even though the software was believed to be open source, and even though no one had knowledge of the patent infringement.[80]

This scenario is possible not only for open source software, but also for proprietary software. A software publishing firm may also discover that their products infringe a patent or other intellectual property. However, commercial enterprises may stand in a better position to consider the intellectual property ramifications of their published software than the more diffuse open source community. Such enterprises often perform audits or establish procedures to avoid the use of other's software in their own products.[81] Members of the more loosely organized open source community may have less capability to engage in these sorts of efforts.

At least one member of the open source community has addressed possible patent issues with respect to open source software. Red Hat, Inc., a well-known distributor of the open source operating system program called Linux, has presented a "patent promise." Red Hat's promise states in part:

Subject to any qualifications or limitations stated herein, to the extent any party exercises a Patent Right with respect to Open Source/Free Software which reads on any claim of any patent held by Red Hat, Red Hat agrees to refrain from enforcing the infringed patent against such party for such exercise . . . .[82]

This statement appears to immunize users of open source software from claims of patent infringement by Red Hat, subject to certain qualifications. Red Hat's promise expressly does not cover patents owned by anyone else, however. As a result, users of open source software theoretically face the possibility of infringement claims by anyone holding a pertinent patent.

The ongoing litigation between the SCO Group and IBM Corporation offers a notable example of third party infringement claims. SCO is the current owner of the source code, as well as certain intellectual property rights, associated with a computer program known as UNIX.[83] UNIX, a widely used operating system program, coordinates use of the computer's resources (such as its disk drive or a printer) during the computer's operation.[84] Publishers of certain versions of UNIX have declared their programs to be open source software.[85]

On March 6, 2003, SCO filed a lawsuit against IBM, in part asserting that IBM had misappropriated SCO's trade secrets.[86] More specifically, SCO contends that IBM accessed its proprietary information when it licensed the UNIX software code from SCO and its predecessors.[87] SCO further asserts that IBM then introduced this proprietary UNIX code into its own operating system, AIX, and later into its Linux-based products.[88] IBM has denied these charges and, in turn, has asserted that SCO products infringe IBM copyrights and patents.[89] As of early 2004, two other open source software vendors, Novell and Red Hat, had also become party to that litigation or to related lawsuits.[90]

In addition to commencing infringement litigation against IBM, SCO has reportedly sent 1,500 letters to other firms. These letters are said to explain that the recipient's use of Linux could expose them to liability, and also to extend an offer of a license.[91] The requested fee is reportedly about $700 for each computer using the Linux code.[92] This demand has in turn animated some observers, such as the Electronic Frontier Foundation, to organize protests against SCO.[93] As well, a consortium of technology companies, including IBM and Intel, have contributed substantial amounts of money towards the "Linux Legal Defense Fund."[94] Contributors intend that these sums be used to defray the legal expenses of open source software users who face charges of intellectual property infringement.[95]

As of the publication date of this report, the lawsuit between SCO and IBM continues in the U.S. District Court for the District of Utah in Salt Lake City. The outcome of this litigation may significantly impact users of Linux and other open source software. It also illustrates the potential tension between intellectual property rights and the open source community. Regardless of the outcome of this particular case, the SCO-IBM dispute demonstrates the possibility of future intellectual property infringement claims against open source software, commenced by individuals who claim not to be subject to the license under which the software was distributed.

Validity of Open Source Licenses

No court has yet ruled on the enforceability of open source software licenses.[96] Some observers have suggested that these agreements may be invalid, however.[97] These commentators point to the doctrine of federal preemption, which invalidates state laws that are inconsistent with federal laws.[98] Under this view, open source software licenses — which are enforced through the mechanisms of state contract laws — conflict with the federal copyright statute, and are therefore invalid.[99]

The rule of federal preemption derives from the Supremacy Clause of the U.S. Constitution. That provision states that the "Constitution and the laws of the United States...shall be the supreme law of the land . . .anything in the constitutions or laws of any State to the contrary notwithstanding."[100] In resolving cases under the Supremacy Clause, courts typically determine whether enforcement of a state law would either directly conflict with federal law or frustrate federal purposes.[101] In addition, section 301 of the Copyright Act expressly exempts state laws "that are equivalent to any of the exclusive rights within the general scope of copyright" and that apply to "the types of works protected by the Copyright Act."[102]

The ongoing litigation between SCO and IBM may offer a court the opportunity to decide whether the open source license known as the General Public License, or GPL, is preempted by the federal copyright law.[103] In that litigation, IBM has in part argued that SCO cannot enforce whatever intellectual property rights SCO owns in Linux due to the terms of the GPL. According to IBM, SCO distributed Linux software under the GPL for many years. IBM observes that the GPL stipulates that software distributed under the GPL must be made available for copying by others. As a result, IBM has asserted that SCO cannot now demand payment for use of any intellectual property SCO might own in Linux.[104]

In turn, SCO has in part asserted that the GPL is invalid. As part of its argument, SCO has reportedly pointed to section 117 of the Copyright Act.[105] That statute provides that "it is not [a copyright] infringement for the owner of a copy of a computer program to make or authorize the making of another copy or adaptation of that computer program provided . . .

that such new copy or adaptation is for archival purposes only . . . ."[106] Stated differently, the Copyright Act permits users to make a backup copy of their software without fear of infringement liability.

The GPL places limitations upon the ability of users to make copies of GPL-licensed software, however. In particular, the GPL requires that individuals receiving copies of the software also receive copies of the GPL, and that persons making copies of the software be able to access the program's source code.[107] The GPL then states that "[a]ny attempt otherwise to copy, modify, sublicense or distribute the Program is void, and will automatically terminate your rights under this License."[108] SCO has taken the position that because the GPL conflicts with the Copyright Act, the GPL is preempted and therefore unenforceable.[109]

Notably, SCO and IBM have presented a number of other arguments to the court. The court may not need to address this issue to resolve their dispute.[110] This aspect of the SCO-IBM litigation has nonetheless raised concerns in the open source software community. As explained by journalist William M. Bulkeley:

> If SCO's argument ultimately wins, free-software advocates worry it would create considerable uncertainty about the legal status of many industry products. Although most believe the industry could adapt, companies that use GPL-licensed software might be confronted by surprise copyright claims from software developers. In addition, creation of new programs might be slowed by the confusion.[111]

Other observers believe that the possibility of a court striking down the GPL is small. For example, lawyer James Boyle, a member of the faculty of the Duke University Law School, has reportedly described the federal preemption argument as "simply ludicrous."[112] Eben Moglen, a law professor at Columbia University, believes that the federal preemption argument is overly broad. In Moglen's view, this position would effectively invalidate all software licenses, even those for which people pay, and even for proprietary software.[113] Moglen believes it unlikely that a court would wish to dispute this longstanding industry practice.[114]

POSSIBLE LEGISLATIVE ISSUES AND APPROACHES

Given the wide recognition that intellectual property and the open source movement are of growing importance in the U.S. computer industry,[115] the relationship between these fields is the subject of increasing attention. The policymakers of the 108th Congress have addressed the open source movement with respect to cybersecurity and other contexts. Should Congress choose to address this area directly, a variety of approaches are available. If the current interface between intellectual property rights and the open source movement is considered satisfactory, then no action need be taken. Indeed, growing awareness that intellectual property and open source software licenses can sometimes conflict may lead to more sophisticated treatment of intellectual property by members of the open source community, as well as continued refinement of the governing law in the courts.

Another approach is to provide governmental assistance to the open source movement in identifying intellectual properties that might bear upon a particular open source software product. For example, the U.S. Patent and Trademark Office could, upon request by a recognized open source software publisher or organization, conduct a search of pending patent applications and issued patents in order to determine whether these patents might bear upon a particular open source software program. This capability would allow members of the open source community to become more fully informed of intellectual property rights. It should be noted, however, that a number of patent research firms already exist that could conduct such a search for a fee, at least with respect to issued patents and published patent applications.[116]

More far-reaching legal reforms are also possible. For example, one recognized source of legal uncertainty for the software industry concerns the enforceability of open source licenses.[117] A legislative statement concerning the status of these licenses in terms of the federal preemption doctrine might allow software firms to make decisions concerning research, investment, and other commercial activities with more confidence.

The allegedly viral nature of open source software presents another source of concern. One possible legislative response is to allow a proprietary software publisher that discovers its product contains an open source component a fixed period of time to eliminate the open source component. If the publisher removes the open source component within the stipulated grace period, then the software would remain proprietary.

Any possible legal reform would be well-advised to recognize that the U.S. software industry is increasingly characterized both by rapid innovation and by a distinct community of knowledgeable users who wish to "opt out" of the intellectual property system. The possibility of intellectual property rights, and their attendant license fees and royalties, may provide a significant incentive for firms to innovate and to distribute software. On the other hand, some computer users believe that these incentives are unnecessary, and further hope to maintain a non-proprietary environment of software distribution and development.[118] These two trends have sometimes led to conflicts between exclusive intellectual property rights and open source software. Striking a balance between promoting innovation, on one hand, and accommodating the demands of software developers and users, on the other, forms an important component of contemporary software policy.

REFERENCES

[1] "Source code" is a set of programming instructions that is typically written in a "high-level" language, such as C or Pascal, that resembles the English language. After a programmer drafts the source code, it is compiled into a machine language that consists entirely of numbers and can be recognized by a computer. Proprietary software is distributed only with the machine code, that allows the computer to function, but cannot be altered by the user. For more on this technology, *see infra* notes 9-12; CRS Report RL31627, *Computer Software and Open Source Issues: A Primer*, by Jeffrey W. Seifert.

[2] Shawn W. Potter, "Opening Up to Open Source," 6 *Richmond Journal of Law and Technology* (2000), 24.

[3] David S. Evans & Bernard J. Reddy, "Government Preferences for Promoting Open-Source Software: A Solution in Search of a Problem," 9 *Michigan Telecommunications & Technology Law Review* (2003), 313.

[4] James A. Harvey & Todd S. McClelland, "*SCO v. IBM:* The Open Source Benefits and Risks Are Real," 20 *Computer & Internet Lawyer* (Sept. 2003), no. 9 at 1.
[5] Natasha T. Horne, "Open Source Software Licensing: Using Copyright Law to Encourage Free Use," 17 *Georgia State University Law Review* (2001), 863.
[6] Christian H. Nadan, "Open Source Licensing: Virus or Virtue?," 10 *Texas Intellectual Property Law Journal* (2002), 349.
[7] Dennis M. Kennedy, "A Primer on Open Source Licensing Legal Issues: Copyright, Copyleft and the Future," 20 *St. Louis University Public Law Review* (2001), 345.
[8] Daniel Lin et al., "Source Code Versus Object Code: Patent Implications for the Open Source Community," 18 *Santa Clara Computer and High Technology Law Journal* (2002), 235.
[9] *Ibid.*
[10] Nadan, *supra* note 6.
[11] *See infra* notes 31-62 and accompanying text.
[12] *See* Patrick K. Bobko, "Linux and General Public Licenses: Can Copyright Keep 'Open Source' Software Free?," 28 *American Intellectual Property Law Association Quarterly Journal* (2000), 81.
[13] Richard Stallman, The GNU Project, at http://www.fsf.org/gnu/thegnuproject.html.
[14] *Ibid.*
[15] R. Polk Wagner, "Information Wants to Be Free: Intellectual Property and the Mythologies of Control," 103 *Columbia Law Review* (2003), 995.
[16] Klaus M. Schmidt & Monika Schnitzer, "Public Subsidies for Open Source? Some Economic Policy Issues," 16 *Harvard Journal of Law and Technology* (2003), 473.
[17] Jason B. Wacha, "Open Source, Free Software, and the General Public License," 20 *Computer and Internet Lawyer* no.3 (March 2003), 20.
[18] Joseph Scott Miller, "Allchin's Folly: Exploding Some Myths About Open Source Software," 20 *Cardozo Arts and Entertainment Law Journal* (2002), 491.
[19] *See* Free Software Foundation, "What is Copyleft?," available at http://www.gnu.org/licenses/licenses.html#WhatIsCopyleft
[20] Kennedy, *supra* note 7.
[21] *See* Open Source Initiative, Introductory Statement (2004), available at http://www.opensource.org.
[22] *See* Open Source Initiative, The Open Source Definition, Version 1.9 (2004), available at http://www.opensource.org/docs/definition.php.
[23] *See* Open Source Initiative, OSI Certification Mark and Program (2004), available at http://www.opensource.org/docs/certification_mark.php.
[24] The Open Source Definition, *supra* note 22.
[25] *See* Terry J. Ilardi, "Mass Licensing," 747 *Practising Law Institute Patents & High Technology Licensing* (2003), 187.
[26] The Free Software Foundation, "The Free Software Definition" (2004), available at http://www.gnu.org/philosophy/free-sw.html.
[27] *Ibid.*
[28] Daniel B. Ravicher, "Facilitating Collaborative Software Development: The Enforceability of Mass-Market Public Software Licenses," 5 *Virginia Journal of Law and Technology* (2000), 11.
[29] Kennedy, *supra* note 7.

[30] David McGowan, "Legal Implications of Open-Source Software," 2001 *University of Illinois Law Review* (2001), 241.
[31] Roger E. Schechter & John R. Thomas, *Intellectual Property: The Law of Copyrights, Patents and Trademarks* (2003), 1-2.
[32] Gordon U. Sanford, III, "An Intellectual Property Roadmap: The Business Lawyer's Role in the Realm of Intellectual Property," 19 *Mississippi College Law Review* (1998), 177.
[33] Simone Rose, "Patent 'Monopolyphobia': A Means of Extinguishing the Fountainhead?," 49 *Case Western Reserve Law Review* 509 (1999).
[34] Jonathan Eaton & Samuel J. Kortum, "Trade in Ideas: Patenting and Productivity in the OECD," 40 *Journal of International Economics* (1996), 251.
[35] Robert P. Merges, "Intellectual Property and the Costs of Commercial Exchange: A Review Essay," 93 *Michigan Law Review* (1995), 1570.
[36] Wendy J. Gordon, "A Property Right in Self Expression: Equality and Individuality in the Natural Law of Intellectual Property," 102 *Yale Law Journal* (1993), 1533.
[37] *See* Justin Hughes, "The Philosophy of Intellectual Property," 77 *Georgetown Law Journal* (1988), 287.
[38] David D. Friedman, et all, "Some Economics of Trade Secret Law," 5 *Journal of Economic Perspectives* (1991), 61.
[39] John R. Thomas, "Collusion and Collective Action in the Patent System: A Proposal for Patent Bounties," *University of Illinois Law Review* (2001), 305.
[40] Mark A. Haynes, "Commentary: Black Holes of Innovation in the Software Arts," 14 *Berkeley Technology Law Journal* (1999), 567.
[41] John Perry Barlow, "The Economy of Ideas," *Wired* 2.03 (March 1994).
[42] *See* F. Scott Kieff, "Property Rights and Property Rules for Commercializing Inventions," 85 *Minnesota Law Review* (2001), 697.
[43] The 1976 Copyright Act, as amended, is codified at 17 U.S.C. § 101 (2000) *et seq.*
[44] 17 U.S.C. §§ 408 - 412 (2000).
[45] 17 U.S.C. § § 401 - 406 (2000).
[46] 17 U.S.C. § 201 (2000).
[47] 17 U.S.C. § 106 (2000).
[48] 17 U.S.C. § 302(a) (2000).
[49] 17 U.S.C. § 201(d) (2000).
[50] The 1952 Patent Act, as amended, is codified in 35 U.S.C. § 1 (2000) *et seq.*
[51] 35 U.S.C. § 131 (2000).
[52] 35 U.S.C. § 112 (2000).
[53] *Ibid.*
[54] 35 U.S.C. § 101. (2000).
[55] 35 U.S.C. § 102 (2000).
[56] 35 U.S.C. § 103 (2000).
[57] 35 U.S.C. § 271(a) (2000).
[58] 35 U.S.C. § 154(a)(2) (2000). Although patent term is based upon the filing date, the patentee gains no enforceable legal rights until the USPTO allows the application to issue as a granted patent. A number of Patent Act provisions may modify the basic 20-year term, considering examination delays at the USPTO and delays in obtaining marketing approval for the patented invention from other federal agencies.

[59] American Law Institute, Restatement of Unfair Competition Third § 39 (1995).
[60] *See* United States v. Dubilier Condenser Corp., 289 U.S. 178, 186, 53 S.Ct. 554, 77 L.Ed. 1114 (1933).
[61] Restatement of Unfair Competition, *supra* note 59, at § 43.
[62] *Ibid* § 39 cmt. c.
[63] Stephen M. McJohn, "The Paradoxes of Free Software," 9 *George Mason Law Review* (2000), 25.
[64] Nadan, *supra* note 6.
[65] Robert W. Gomulkiewicz, "How Copyleft Uses License Rights to Succeed in the Open Source Software Revolution and the Implications for Article 2B," 36 *Houston Law Review* (1999), 179.
[66] Nadan, *supra* note 6.
[67] Thomas M. Pitegoff, "Open Source, Open World: New Possibilities for Software in Business," 11 *Business Law Today* (Sept.-Oct. 2001), 52.
[68] *See supra* notes 25-27 and accompanying text.
[69] *See* Free Software Foundation, *Frequently Asked Questions about the GNU GPL*, available at http://www.gnu.org/licenses/gpl-faq.html#GPLInProprietarySystem.
[70] *Ibid.*
[71] *Ibid.*
[72] Seifert, *supra* note 1.
[73] Harvey & McClelland, *supra* note 4.
[74] Miller, *supra* note 18.
[75] Jonathon Krim, "Open-Source Fight Flares at the Pentagon; Microsoft Lobbies Hard Against Free Software," *Washington Post* (May 23, 2002), E1.
[76] Ravicher, *supra* note 28.
[77] *See supra* notes 28-30 and accompanying text.
[78] Ira V. Heffan, "Copyleft: Licensing Collaborative Works in the Digital Age," 49 *Stanford Law Review* (1997), 1487.
[79] Harvey & McClelland, *supra* note 4.
[80] *Ibid.*
[81] *Ibid.*
[82] Red Hat, Inc., Statement of Position and Our Promise on Software Patents, http://www.redhat.com/legal/patent_policy.html (2004).
[83] The term "UNIX" is a trademark registered to the Open Group. *See* http://www.opengroup.org.
[84] *See* Lucent Laboratories, The Creation of the UNIX Operating System (available at http://www.bell-labs.com/history/unix/).
[85] *See* Joel West, "How open is open enough? Melding proprietary and open source platform strategies," 32 *Research Policy* no. 7 (July 2003), 1259.
[86] Harvey & McClelland, *supra* note 4.
[87] *Ibid.*
[88] Reuters News, SCO Advances Suit Against IBM (June 29, 2003).
[89] Robert McMillan, "SCO infringed copyright, IBM alleges in new claim," *Computer world* (Sept. 26, 2003).
[90] ComputerWire News, "SCO Hits Novell with IP Slander Action: (Jan. 21, 2004).

[91] David Bank, "Lawsuit Over Copyrights Is Part of Effort to Collect License Fees for Software," *The Wall Street Journal* (Nov. 19, 2003).
[92] *Ibid.*
[93] *See* Electronic Frontier Foundation, "Linux Users Unite: Stop SCO!," (2004), available at http://action.eff.org/action/index.asp?step=2&item=2775.
[94] Dean Takahashi, "Aid group to defend Linux suits: Software Backers' Aim is $10 Million," *San Jose Mercury News* (Jan. 13, 2004), 1.
[95] Keith Regan, "IBM, Intel Back Linux Legal Defense Fund," *E-Commerce Times* (Jan. 22, 2004) (available at http://www.ecommercetimes.com/perl/story/32570.html).
[96] *See* Ravicher, *supra* note 28.
[97] *See* McGowan, *supra* note 30.
[98] Paul Heald, "Federal Intellectual Property Law and the Economics of Preemption," 76 *Iowa Law Review* (1991), 959.
[99] William M. Bulkeley, "Linux Lawsuit Could Undercut Other 'Freeware'," *The Wall Street Journal* (Aug. 14, 2003), B1.
[100] U.S. Constitution, Article VI.
[101] Heald, *supra* note 98.
[102] 17 U.S.C. § 301 (2000).
[103] For an introduction to the litigation between SCO and IBM, see *supra* notes 83-95. The GPL is discussed *supra* at notes 25-27.
[104] Bulkeley, *supra* note 99.
[105] 17 U.S.C. § 117 (2000); *see supra* notes 43-49 and accompanying text.
[106] 17 U.S.C. § 117 (2000).
[107] GNU General Public License, Version 2 (June 1991), available at http://www.gnu.org/licenses/gpl.txt.
[108] *Ibid.*
[109] Bulkeley, *supra* note 99.
[110] *Ibid.*
[111] *Ibid.*
[112] Evan P. Schultz, "A Brewing Legal Battle Imperils Linux and Other Public-License Software," 26 *Legal Times IP Magazine* no. 41 (Oct. 13, 2003), 14.
[113] *Ibid.*
[114] *Ibid.*
[115] Mathias Strasser, "A New Paradigm in Intellectual Property Law? The Case Against Open Sources," *Stanford Technology Law Review* (2001), 4.
[116] Steve D. Beyer, "Searching — The Art Behind An Opinion," 667 *Practising Law Institute Patents, Copyrights, Trademarks and Literary Property Course Handbook Series* (Nov. 1, 2001), 45.
[117] *See supra* notes 96-114 and accompanying text.
[118] *See* Andrea Bonaccorsi & Cristina Rossi, "Why Open Source software can succeed," 32 *Research Policy* no. 7 (July 2003), 1243.

In: New Intellectual Property Issues
Editor: John V. Luong, pp. 75-95
ISBN: 978-1-60021-211-6
© 2007 Nova Science Publishers, Inc.

Chapter 4

INTELLECTUAL PROPERTY AND THE FREE TRADE AGREEMENTS: INNOVATION POLICY ISSUES*

John R. Thomas

SUMMARY

The United States has recently entered into a number of Free Trade Agreements, or FTAs, with Australia, Chile, Singapore, and other trading partners. Negotiations are currently ongoing with respect to the establishment of additional FTAs. In keeping with a congressional directive established in the Bipartisan Trade Promotion Act of 2002, P.L. 107-210, one objective of forming the FTAs is to establish "a standard of [intellectual property] protection similar to that found in United States law." As a result, most of the FTAs stipulate minimum levels of protection with respect to copyrights, data protection, patents, trademarks, and other forms of intellectual property. These standards relate to such provisions as the term of protection, scope of rights, and mechanisms by which these intellectual property rights are acquired and enforced.

The different FTAs vary in their comprehensiveness and level of detail. Each of these agreements has nonetheless been drafted in a manner that complies with current U.S. law. As a result, the effect of each FTA is to obligate signatories to such agreements to amend their intellectual property laws to match or resemble those of the United States.

The FTAs have been described as an effective mechanism for advancing U.S. interests in securing intellectual property protection. Increased levels of intellectual property protection with respect to computer software, music, motion pictures, and pharmaceuticals may promote a more favorable balance of trade for U.S. industry, decrease domestic prices for innovative goods and services, and serve other policy goals. The FTA framework has at times proven to be a more advantageous forum for achieving the intellectual property goals of the United States than multilateral settings.

Nonetheless, concerns have arisen over the intellectual property provisions of the FTAs. Some observers believe that certain FTA provisions may lock the United States into current intellectual property policies, inhibiting opportunities for future reform. Other commentators are concerned that under existing multilateral agreements, in particular those of the World Trade Organization, the intellectual property obligations found within one FTA may extend beyond that particular treaty partner. Finally, some observers perceive the FTAs to be an inappropriate and unfair vehicle for international intellectual property reforms due to the

* Excerpted from CRS Report RL33205, Updated December 21, 2005

strong bargaining position of the United States. The scope of these potential consequences counsels continued congressional attention towards the FTAs that the United States has already formed, as well as those FTAs that are planned for the future.

Congressional interest in the intellectual property laws, including patents, copyrights, and trademarks, has been reflected in recent Free Trade Agreements (FTAs) to which the United States is a signatory. Congress stated in the Bipartisan Trade Promotion Act of 2002[1] that an overall negotiating objective of such agreements was to encourage our treaty partners to agree to "a standard of [intellectual property] protection similar to that found in United States law."[2] In keeping with this mandate, the United States has entered into numerous FTAs that have required their signatories to conform to stipulated standards of intellectual property protection.[3] Because the FTAs are drafted in a manner that complies with current U.S. law, their effect is to obligate U.S. treaty partners to amend their intellectual property laws to match or resemble those of the United States.[4]

FTAs have been described as an effective mechanism for advancing U.S. interests in ensuring intellectual property protection.[5] Nonetheless, concerns have arisen over the intellectual property provisions of the FTAs. Some commentators have suggested that certain FTA provisions may lock the United States into current intellectual property policies, inhibiting opportunities for future reform.[6] Others are concerned that under existing multilateral agreements, in particular those of the World Trade Organization (WTO), the intellectual property obligations found within one FTA may extend beyond that particular treaty partner.[7] Finally, some observers perceive the FTAs to be an inappropriate and unfair vehicle for international intellectual property reforms.[8]

This report offers a broad overview of the intellectual property components of the FTAs. It begins by offering a brief introduction to the global intellectual property system. Next, this report considers the subject matter of the different free trade agreements themselves. It then reviews perceived concerns with respect to the free trade agreements and closes with concluding observations.

Intellectual Property Fundamentals

The term "intellectual property" refers to the subject matter of the laws that give rise to proprietary interests in creations of the mind. Significant legal intellectual property disciplines include copyright, which concerns artistic and literary works; patent, pertaining to pragmatic innovations; trademark, relating to commercial symbols; and data protection, concerning pharmaceutical and agricultural test data.[9] A brief review of each of these legal regimes, as they exist in the United States, will aid subsequent analysis.

Copyrights provide protection for original works of authorship. The types of creations addressed by copyright range from traditional works of art, including literature, music and visual art, to such modern forms of artistic expression as sound recordings, motion pictures and even computer software.[10] Copyright protection arises automatically, as soon as the work has been fixed in tangible form.[11] Authors may register their works with the U.S. Copyright Office, a component of the Library of Congress. If an author chooses to register his work, he obtains certain procedural and substantive advantages during copyright enforcement.[12] The copyright law affords authors the exclusive right to reproduce, adapt, and publicly distribute, perform, and display the protected work, subject to limitations such as the fair use privilege.[13] The term of copyright is ordinarily the life of the author plus 70 years.[14]

Patents provide exclusive rights to inventors of new, useful and nonobvious inventions.[15] The patent law concerns hard technologies, including chemical, electrical and mechanical products and processes, as well as other pragmatic innovations in fields ranging from biotechnology to business methods.[16] An inventor may obtain a patent by filing a patent application with the United States Patent and Trademark Office (USPTO). Such an application must completely describe and precisely claim the invention.[17] Patents issued confer the right to exclude others from making, using, selling, offering to sell or importing into the United States the patented invention.[18] The term of a patent is ordinarily 20 years from the date the application was filed.[19]

Trademarks consist of any word or symbol used by a merchant to identify its goods or services, and to distinguish them from those of others.[20] To be subject to protection under the trademark laws, a mark must successfully distinguish the origins of its associated goods, and not be confusingly similar to marks used by others or merely describe the characteristics of those goods.[21] Trademark rights arise under state law as soon as the mark is used on goods in commerce.[22] However, trademarks may be registered with the USPTO, a step that affords significant substantive and procedural advantages.[23] Trademark law also protects the appearance of product packaging and, in some cases, the actual physical configuration of the goods, if these serve as brand identifiers. A trademark owner may prevent others from using any mark that creates a likelihood of confusion as to the source or sponsorship of the associated goods or services.[24] Trademark rights persist so long as the mark continues to be used and retains its distinctiveness.[25]

Another intellectual property right concerns the protection of pharmaceutical and agricultural chemical test data. Most national governments regulate the marketing of pharmaceuticals and agricultural chemicals in the interest of public health.[26] This oversight includes a required showing that these products are safe and effective before they can be distributed to the public. Such evidence is not generated by government laboratories, however. Rather, national governments require sponsors of new drugs and agricultural chemicals to submit test data that evidence their safety and efficacy.[27]

Because the required test data may be very costly to generate, sponsors of new pharmaceuticals and agricultural chemicals often do not disclose test data to the public. Otherwise the sponsor's competitors could avoid the expenses of developing this data for themselves. Sponsors also desire government agencies that receive the test data to maintain the data in confidence.[28] In addition, brand-name pharmaceutical and agricultural chemical firms often seek to limit the ability of generic competitors to rely upon the data in order to obtain marketing approval of their own products, even if they cannot actually view the data themselves.[29]

In the United States, federal legislation limits the ability of generic competitors to reference data generated by the manufacturers of brand-name pharmaceuticals and agricultural chemicals. With respect to pharmaceuticals, these "marketing exclusivities" consist of a period of time during which the Food and Drug Administration (FDA) protects an approved drug protection from competing applications for marketing approval. Under the Hatch-Waxman Act,[30] drugs that qualify as a "new chemical entity"[31] may obtain five years of marketing exclusivity.[32] The Hatch-Waxman Act also provides for a three-year marketing exclusivity awarded to a drug sponsor that conducts new clinical studies upon a drug that does not qualify as a new chemical entity.[33] The Federal Insecticide, Fungicide, and Rodenticide Act of 1978,[34] also provides for data protection for agricultural chemicals under somewhat different certain terms and conditions.[35]

A number of rationales explain the intellectual property laws. Some legal experts assert that intellectual property laws are needed to encourage individuals to create new works of authorship, inventions, and other innovative subject matter. It is believed that absent a system of proprietary rights, free riders could readily exploit these works without having to bear the costs of creating them. Individuals would in turn possess diminished incentives to devote their efforts to being authors or inventors.[36] In addition, intellectual property rights are said to facilitate market mechanisms by creating discrete, well-defined property interests that decrease transaction costs and encourage commercial exchanges.[37]

On the other hand, intellectual property laws have been subject to criticism. Some assert that intellectual property rights are unnecessary due to market forces that already suffice to create an optimal level of creative activity. The desire to become famous, or to gain a lead time advantage over competitors, may well provide sufficient inducement to write or to invent without the need for further incentives.[38] In an era where information can be readily transmitted around the globe, the notion that an innovation can be an object of possession has also been challenged.[39]

Each of these contentions has a varying degree of persuasiveness. As of yet, however, no conclusive study has demonstrated that the social benefits of the intellectual property laws outweigh the costs. As a result, the justifications and criticisms of intellectual property remain open to challenge.[40]

The TRIPS Agreement

At present time, no unified legal regime governing intellectual property rights exists on an international basis. There is no global copyright, patent or trademark. Innovators must secure and enforce these rights within the particular jurisdiction where they desire protection.[41]

In addition, the varying national laws that establish intellectual property rights often differ in important respects. Such factors as the works subject to protection, the scope of rights, and the duration of protection are among the factors that may vary widely in different jurisdictions.[42] Absent some obligation, a particular nation is not required to provide intellectual property rights that resemble those of the United States, or even to establish such laws at all.[43]

Since the nineteenth century, however, many nations in fact have committed to provide certain standards of intellectual property protection via bilateral and multilateral international agreements.[44] The most significant of these is the WTO Agreement on Trade — Related Aspects of Intellectual Property Rights, commonly known as the TRIPS Agreement.[45] Domestically, the Uruguay Round Agreements Act in 1995 made significant amendments to U.S. intellectual property laws in an effort to comply with the TRIPS Agreement.[46] As discussed below, the TRIPS Agreement required each WTO member state to provide minimum substantive standards of intellectual property protection and enforcement.

General Provisions

The TRIPS Agreement requires each WTO member state to observe the standards of "national treatment" and "most-favored-nation" with respect to its intellectual property laws.[47] Following the national treatment principle, WTO members agree to treat foreign inventors no worse than domestic inventors in their patent laws, so long as these foreign inventors are nationals of a WTO member state. It would be impermissible, for example, for the USPTO to charge nationals of a WTO member state a higher application fee than is required of U.S. citizens, or to provide a shorter patent term for such inventors than is granted to U.S. inventors. Under the most favored nation provision, with limited exceptions, any privilege granted to nationals of one WTO member state must also be afforded to nationals of all WTO member states.[48]

The TRIPS Agreement also requires each WTO member state to comply with certain provisions of earlier international agreements pertaining to intellectual property.[49] These treaties include the Berne Convention,[50] which relates to copyrights; the Paris Convention,[51] which concerns patents and trademarks; and the Treaty on Intellectual Property in Respect of Integrated Circuits,[52] concerning semiconductor chip topography designs.

Copyrights

The TRIPS Agreement obligates each WTO member state to provide a term of copyright protection of no less than 50 years from the death of the author.[53] The TRIPS Agreement additionally requires WTO member states to protect computer programs as literary works.[54] Protection must also be accorded to data compilations that, by virtue of their selection or arrangement, constitute "intellectual creations."[55] The TRIPS Agreement also calls for "rental rights" — the right to authorize or prohibit the commercial rental to the public of a protected work of authorship — with respect to computer programs and cinematographic works.[56]

Patents

Following the TRIPS Agreement, each WTO member state agreed to allow patents to issue on inventions "in all fields of technology, provided that they are new, involve an inventive step and are capable of industrial application."[57] The TRIPS Agreement includes some exceptions to this broad principle, however. Certain methods of medical treatment, plants and animals other than microorganisms, and inventions that violate the "*ordre public* or morality" may be excluded from patentability at the option of the member state.[58]

WTO member states also agreed that patentees shall have the right to exclude others from making, using, offering for sale, selling, or importing the patented invention,[59] subject to limited exceptions.[60] The TRIPS Agreement further stipulates that the term of patent protection available shall not end before the expiration of a period of 20 years counted from the filing date.[61]

The TRIPS Agreement places some limits upon the ability of WTO member states to award compulsory licenses for the use of another's patented invention.[62] Among the most detailed provisions of the TRIPS Agreement, Article 31 in part imposes the following restrictions upon the issuance of compulsory licenses:

- Each application for a compulsory license must be considered on its individual merits.
- The proposed user must have made efforts to obtain authorization from the patent owner on reasonable commercial terms and conditions and must demonstrate that such efforts have not been successful within a reasonable period of time. However, this requirement may be waived in the case of a national emergency or other circumstances of extreme urgency.
- The compulsory license must be revocable if and when its motivating circumstances cease to exist and are unlikely to recur.
- The patent owner must be paid adequate remuneration in the circumstances of each case, taking into account the economic value of the authorization.
- The legal validity of any decision relating to the authorization of such use shall be subject to judicial or other independent review.
- Any such use shall be authorized predominantly for the supply of the domestic market of the member authorizing such use.

Many nations considered the last of the restrictions noted above — requiring that any compulsory license be authorized predominantly for local use — to be a burdensome standard. Some countries may lack the technological or financial capabilities to manufacture advanced products, including certain pharmaceuticals needed to combat AIDS or other epidemics. The WTO ministerial conference held in Doha, Qatar in 2001 recognized this concern when it issued a "Declaration on the TRIPS Agreement and Public Health."[63] The result was a 2003 decision of the WTO General Council that limited the domestic supply requirement with respect to pharmaceuticals needed to address public health problems.[64] Under the 2003 decision, any least developed country, as well as any other country that certifies it has insufficient manufacturing capabilities to manufacture a patented drug, may issue a compulsory license that allows for domestic public health needs to be satisfied through importation.[65] On December 6, 2005, WTO members agreed to incorporate the language of the 2003 decision directly within the TRIPS Agreement itself.[66]

Trademarks

The TRIPS Agreement provides that "[a]ny sign, or combination of signs, capable of distinguishing the goods or services of one undertaking from those of other undertakings, shall be capable of constituting a trademark." The TRIPS Agreement further requires that a trademark owner, with some exceptions,[67] "shall have the exclusive right to prevent all third parties not having the owner's consent from using in the course of trade identical or similar signs for goods or services which are identical or similar to those in respect of which the trademark is registered where such use would result in a likelihood of confusion."[68]

Trademarks that have become well-known in a particular country are entitled to additional protection.[69]

Trademarks are subject to a minimum term of 7 years, which may be renewed indefinitely.[70] The TRIPS Agreement further stipulates that compulsory licensing of trademarks shall not be permitted.[71] Under this rule, legal requirements that foreign marks be used in conjunction with local marks are, as a general matter, prohibited.

Data Protection

The TRIPS Agreement requires each WTO member state to establish protections for pharmaceutical and agricultural chemical test data under certain conditions. Although this obligation is stated succinctly in Article 39.3 of the TRIPS Agreement, it has proven controversial. Article 39.3 specifically provides:

> Members, when requiring as a condition of approving the marketing of pharmaceutical or of agricultural chemical products which utilize new chemical entities, the submission of undisclosed test or other data, the origination of which involves a considerable effort, shall protect such data against unfair commercial use. In addition, Members shall protect such data against disclosure, except where necessary to protect the public, or unless steps are taken to ensure that the data are protected against unfair commercial use.

Many commentators have observed that, in contrast to the more specific wordings of many other TRIPS Agreements provisions, Article 39.3 establishes broad parameters using vague language.[72] In particular, terms such as "new chemical entities," "considerable effort," and "unfair commercial use" receive no further definition within the TRIPS Agreement. As a result, the precise nature of the obligations Article 39.3 imposes upon WTO member states is not entirely clear.[73]

Other Intellectual Property Rights

The TRIPS Agreement also establishes minimum standards of substantive protection with respect to a number of other intellectual property rights. WTO member states must protect certain geographical indications, which are defined as "indications which identify a good as originating in the territory of a Member, or a region or locality in that territory, where a given quality, reputation or other characteristic of the good is essentially attributable to its geographical origin."[74] They must also protect industrial designs, which consist of a shape, configuration, pattern, or ornament applied to an article manufacture.[75] In addition, they must also provide a cause of action for the misappropriation of trade secrets.[76]

Enforcement

Along with other commitments made by WTO member states, TRIPS Agreement obligations are subject to enforcement through the WTO Dispute Settlement Body (DSB).[77] If one WTO member state believes that another member state is in violation of the TRIPS Agreement, the member states may enter into consultation through the DSB. If the member states cannot resolve their disagreement, the DSB will convene a panel to hear and resolve the dispute. Panel decisions are subject to review by the DSB Appellate Body. The WTO Agreement calls for compensatory trade measures in circumstances where the DSB finds a WTO member state to be in violation of the TRIPS Agreement, yet that member state does not amend its laws. If the parties are unable to agree upon mutually acceptable compensation, then the complaining state may impose limited trade sanctions, such as heightened tariff rates commensurate with the determined injury resulting from the offending practice, against the other member state.[78]

Effective Dates

The various patent portions of the TRIPS Agreement feature a variety of effective dates. These dates depend upon whether the WTO member state designates itself a developed, developing or least developed country. For WTO members other than developing and least developed countries, the compliance date for all requirements of the TRIPS Agreement was set to January 1, 1996.[79]

For signatory states designated as developing countries, the TRIPS Agreement set the general compliance date as January 1, 2000. However, there is one exception to this general date. If on January 1, 2000, a developing country did not extend patent protection to all areas of technology within the meaning of Article 27, that developing country may delay implementation of these provisions for an additional five years. Prior to the TRIPS Agreement, for example, many developing countries did not allow patents to issue on pharmaceuticals. The practical effect of this additional transition period was that developing countries did not need to allow patents concerning pharmaceuticals until January 1, 2005.

Least-developed countries are entitled to a lengthier transition period in implementing TRIPS Agreement obligations. Effective dates with respect to least-developed countries are found in the TRIPS Agreement itself, as well as the 2001 "Declaration on the TRIPS Agreement and Public Health" issued following the Doha Ministerial.[80] Following 2005 modifications to the text of the TRIPS Agreement, a least-developed country may delay implementing the TRIPS Agreement until July 1, 2013. A showing of hardship may qualify least-developed countries for further delays and other concessions.[81] The Doha Declaration further excuses least-developed country members from granting or enforcing patents on pharmaceuticals through January 1, 2016, without prejudice to their ability to seek other extensions of these transitions periods.

The TRIPS Agreement does not oblige its signatories to protect subject matter that fell into the public domain prior to the time its obligations became effective.[82] For example, suppose that a particular developing country (as compared to a developed or least-developed country) traditionally did not allow patents to issue on pharmaceuticals. If that developing country joined the WTO, it must amend its patent law to authorize pharmaceutical patents. The TRIPS Agreement requires only that patents be allowed on new products as of January 1, 2005, however, and does not mandate that patents be granted retroactively. As a result, within that developing country, patent protection need not be afforded to pharmaceuticals that were public available prior to January 1, 2005, even if those pharmaceuticals were patented elsewhere.

Debate on the TRIPS Agreement

The TRIPS Agreement has generated considerable controversy. Some commentators believe that the TRIPS Agreement has led to large transfers of wealth from poor countries to the developed world, and in particular to the United States.[83] Others have contended that the introduction of strong intellectual property rights protection into the developing world restricts sustainable development and perpetuates their dependence upon developed nations.[84] Still others believe that deleterious public health consequences will result from the TRIPS Agreement requirement that patents issue on pharmaceuticals.

Proponents of the TRIPS Agreement instead believe that the introduction of full-fledged intellectual property laws around the globe will provide needed incentives for investment and innovation.[85] Such efforts could promote solutions to problems that are particular to the developing world, including the provision of nutritional needs and cures for diseases not common in the developed world. Supporters also observe that the TRIPS Agreement was one component of a multi-faceted WTO agreement, and believe that the developing world obtained trade benefits in exchange for assuming obligations to protect intellectual property. Even after a decade of experience with the TRIPS Agreement, the exchange of views about possible reforms to the TRIPS Agreement continues at a brisk pace.

Intellectual Property and Free Trade Agreements

The TRIPS Agreement obligated each WTO member state to provide stated levels of intellectual property protection. Although commentators broadly agree that the TRIPS Agreement was a watershed event for international intellectual provision,[86] many of its obligations nonetheless fell short of the level of intellectual property rights available in the United States. For example, U.S. copyright law provides for a term of protection of the life of the author plus 70 years,[87] while a term of life plus 50 years is acceptable under the TRIPS Agreement.[88] U.S. patent law allows patent term to be extended in order to account for long delays in Patent Office acquisition procedures,[89] but the TRIPS Agreement does not require such a measure. The TRIPS Agreement allows WTO member states to disallow the registration of trademarks that are not visually perceivable[90] — an option that blocks protection for such commercial symbols as sounds and scents — but no such blanket restriction exists under U.S. law.

In enacting the Bipartisan Trade Promotion Act of 2002, Congress stated that an overall negotiating objective of such agreements was to encourage our treaty partners to agree to "a standard of [intellectual property] protection similar to that found in United States law."[91] In keeping with this mandate, the United States has entered into numerous FTAs that have required their signatories to provide higher levels of intellectual property protection than are required under the TRIPS Agreement.[92] With respect to the intellectual property laws, each of the FTAs stipulates standards that comply with current U.S. law. As a result, the United States need take no action in order to comply with those provisions of the FTA. In contrast, the U.S. treaty partner must often substantially amend its intellectual property laws in order to comply with the FTA.[93]

This report next offers an overview of these FTA "TRIPS-plus"[94] intellectual property obligations. It should be noted that the intellectual property provisions of most of these agreements are lengthy and detailed. Further, some of the FTAs impose more significant obligations than others.[95] As a result, if particular information about an individual FTA is needed, that instrument should be consulted for a more detailed understanding of its contents.[96]

Copyrights

Several of the FTAs oblige their signatories to provide a copyright term equal to the life of the author plus 70 years, twenty years longer than the standard set by the TRIPS Agreement. Certain FTAs also include obligations that track those established domestically by the Digital Millennium Copyright Act (DMCA) of 1998.[97] In particular, FTA signatories agree to outlaw the circumvention of a technological copyright-control measure, subject to certain exceptions. Descrambling a scrambled work, decrypting an encrypted work, and bypassing a technological measure without the authority of the copyright owner are among the activities that would violate the DMCA. In addition, the FTAs also commonly require their signatories to prohibit the removal or alteration of "rights management information," which is in turn defined to mean electronic information that identifies a protected work and its author, as well as the terms and conditions of the use of the work.

Patents

Several FTAs require their signatories to provide a one-year "grace period" to patent applicants. Under these provisions, individuals are granted one year from the date they publicly disclose their inventions to decide whether to file a patent application or not.[98] FTAs also commonly require their signatories to accede to the Patent Cooperation Treaty,[99] an international agreement that expedites multinational patent acquisition procedures.[100]

Certain of the FTAs also call for patent term extension based upon administrative delays. The TRIPS Agreement obligates all WTO members to provide a patent term of twenty years from the date of filing.[101] However, patent applicants obtain no enforceable rights until such time as the patent actually issues. As a result, extensive Patent Office delays could substantially diminish the period of time in which the patent proprietor enjoys enforceable patent rights. Certain of the FTAs therefore call for patent term extension in the event of "unreasonable delays in a Party's issuance of patents . . . ."[102]

Some FTAs also place additional conditions on the grant of compulsory licenses, beyond the requirements of the TRIPS Agreement. For example the Australia-U.S. FTA provides that a compulsory license may be issued only to remedy an antitrust violation or in cases of public non-commercial use, national emergency, or other circumstances of extreme urgency. In these latter instances, the FTA signatory must limit use of the patented invention to the government or a government-authorized entity and shall provide the patent proprietor with reasonable compensation. Further, the patent proprietor is not required to provide additional undisclosed information or technical know-how that is related to the patented invention.[103]

Trademarks

A number of the FTAs also pertain to the trademark law. They stipulate that a mark need not be visually perceptible in order to be registered as a mark, thereby allowing sounds and scents to serve as marks.[104] The FTAs also stipulate that licenses for trademarks need not be publicly registered to be valid,[105] and provide that disputes over Internet domain names should be resolved through recourse to the principles established in the Uniform Domain Name Dispute Resolution Policy.[106]

Data Protection

Certain FTAs require their signatories to provide five years of marketing exclusivity for pharmaceuticals that utilize new chemical entities. For example, Article 15.10:1(a) of the Dominican Republic-Central America-United States Free Trade Agreement provides:

> If a Party requires, as a condition of approving the marketing of a new pharmaceutical or agricultural chemical product, the submission of undisclosed data concerning safety or efficacy, the Party shall not permit third persons, without the consent of the person who provided the information, to market a product on the basis of (1) the information, or (2) the approval granted to the person who submitted the information for at least five years for pharmaceutical products and ten years for agricultural chemical products from the date of approval in the Party.[107]

The term "new product" is generally defined as "one that does not contain a chemical entity that has been previously approved in the territory of the Party."[108]

Some of the FTAs additionally require their signatories to allow a three-year term of marketing exclusivity in various circumstances for pharmaceuticals that do not qualify as new products.[109] Among them is the Australia-United States Free Trade Agreement, which provides in part:

> With respect to pharmaceutical products, if a Party requires the submission of: (a) new clinical information (other than information related to bioequivalency) or (b) evidence of prior approval of the product in another territory that requires such new information, which is essential to the approval of a pharmaceutical product, the Party shall not permit third persons not having the consent of the person providing the information to market the same or a similar pharmaceutical product on the basis of the marketing approval granted to a person submitting the information for a period of at least three years from the date of the marketing approval by the Party or the other territory, whichever is later.[110]

Other Provisions

The FTAs also address other intellectual property rights. For example, the U.S.-Australia FTA requires both signatories to provide certain levels of protection to "encrypted programme-carrying satellite signals."[111] In addition, the FTAs commonly call for minimal standards concerning litigation proceedings, remedies, and other intellectual property enforcement measures that exceed those of the TRIPS Agreement.[112] The FTAs also set out detailed procedures for the resolution of disputes over compliance with their provisions.[113]

INNOVATION POLICY ISSUES

Observers have acknowledged that the FTAs have been an effective mechanism for advancing U.S. interests in intellectual property on the global stage.[114] Concerns have nonetheless arisen about the implications of the FTA intellectual property provisions. This report considers three of these concerns: potential lock-in effects, extent of FTA obligations, and the fairness of using FTAs as a mechanism for international intellectual property reforms.

U.S. Interests in the FTA Intellectual Property Provisions

The United States has for many years pursued a policy of encouraging our trading partners to expand legal protection for intellectual property.[115] Decreased levels of intellectual property piracy with respect to computer software, music, motion pictures, and pharmaceuticals may promote a more favorable balance of trade for U.S. industry. As Professor J. Thomas McCarthy, a member of the faculty of the University of San Francisco Law School, clearly stated: "We care because if no intellectual property protection exists regarding technical and entertainment information, then we have little to sell the rest of the world."[116]

Enhanced levels of intellectual property protection around the world may also serve other goals of the United States. For example, pharmaceuticals are often sold at higher prices in the United States than in other nations.[117] The previous unavailability of patent protection for innovative pharmaceuticals in some foreign jurisdictions is one factor that may have contributed to this price difference. The introduction of patent protection for pharmaceuticals in developing countries will potentially allow innovative drug companies to recover a portion of their R&D expenses in foreign jurisdictions. In turn, because U.S. consumers will no longer be responsible for funding such a large share of the R&D expenditures of innovative drug companies on a global basis, the price of medications may decrease in the United States. It is possible, however, that developing country markets may not prove sufficiently profitable such that prices would be reduced for U.S. consumers.[118]

In comparison with larger multilateral settings, FTAs may provide a more effective mechanism for advancing the intellectual property interests of the United States. USPTO officials have, for example, expressed frustration over slow progress in consultations over a proposed patent law harmonization treaty before the World Intellectual Property Organization (WIPO).[119] One recent USPTO news release has stated that recent events at WIPO raise "serious questions as to whether WIPO is even a viable forum for further meaningful patent discussions."[120] In contrast, the FTAs have already resulted in significant patent law reforms for many U.S. trading partners. In addition, by establishing certain multinational intellectual property norms with the assistance of a number of trading partners, the United States may be better able to negotiate on a multilateral basis in the future.[121]

Potential Lock-In Effects

Although many observers have cited the merits of FTAs with respect to the intellectual property interests of the United States, others have expressed concerns about the effects or implications of these agreements. Some commentators have expressed concern that the FTAs may potentially introduce complications to U.S. law reform efforts, even in circumstances where modifications to domestic intellectual property rules are considered desirable in the future.[122] In particular, should the Congress later wish to make changes to its laws that are inconsistent with the terms of the FTAs, the United States would either have to renegotiate these agreements, or face the possibility of violating their terms. A potential example of this situation involves the importation of patented pharmaceuticals into the United States, a topic that the Singaporean, Australian, and Moroccan FTAs address. For example, Article 15.9, paragraph 4 of the United States — Morocco FTA provides:

> Each Party shall provide that the exclusive right of the patent owner to prevent importation of a patented product, or a product that results from patented process, without the consent of the patent owner shall not be limited by the sale or distribution of that product outside its territory. [Footnote 10: A Party may limit application of this paragraph to cases where the patent owner has placed restrictions on importation by contract or other means.]

This provision impacts, among other topics, a practice commonly termed "parallel importation"[123] or "reimportation."[124] Parallel imports are authentic products that are legitimately distributed abroad and then sold in the United States, without the permission of the authorized U.S. dealer. These goods are legitimate in that they are produced by the brand-name drug company or its authorized representative. In particular, parallel imports are not generic versions of a brand-name drug distributed by a different manufacturer, nor are they pirated copies that form part of the "black market." Because parallel imports disrupt the marketing arrangements established by the brand-name drug company, however, they are sometimes called "grey market goods."[125]

Under current judicial precedent,[126] U.S. patents may be used to block imports of proprietary pharmaceuticals. The fact that the patent proprietor or its representative previously sold the drug outside the United States does not affect this analysis. This rule rejects a principle termed "international exhaustion," under which a sale of a drug anywhere in the world exhausts the U.S. patent. The FTAs that address this issue therefore comport with governing U.S. judicial opinions.

Proposed legislation before the 109th Congress would alter the current rules concerning the parallel importation of patented pharmaceuticals. The proposed Affordable Health Care Act (S. 16, Sen. Kennedy) would in part introduce the following provision into the patent statute:

> It shall not be an act of infringement to use, offer to sell, or sell within the United States or to import into the United States any patented invention under section 804 of the Federal Food, Drug, and Cosmetic Act that was first sold abroad by or under authority of the owner or licensee of such patent[127]

If this bill, or a similar bill, were passed into law or enacted, the "international exhaustion" principle would apply to foreign sales of patented pharmaceuticals. In circumstances where the patent owner sold the patented pharmaceutical abroad, the patent owner would no longer be able to block imports of these products into the United States. This result appears to conflict with Article 15.9, paragraph 4 of the United States — Morocco FTA, as well as the counterpart provisions of the Singaporean and Australian FTAs.

Perhaps out of recognition of the impact of the FTAs on the parallel importation issue, Rep. Northrup introduced an amendment to the Departments of Commerce and Justice, Science, and Related Appropriations Act, 2006, P.L. 109-108. Section 631 of that bill provides:

None of the funds made available in this Act may be used to include in any new bilateral or multilateral trade agreement the text of—

(1) paragraph 2 of Article 16.7 of the United States-Singapore Free Trade Agreement;
(2) paragraph 4 of Article 17.9 of the United States-Australia Free Trade Agreement; or
(3) paragraph 4 of Article 15.9 of the United States-Morocco Free Trade Agreement.

This provision would apparently seek to prevent future FTAs from incorporating language rejecting an "international exhaustion" rule for pharmaceuticals. This provision does not address the existing Singaporean, Australian, and Moroccan FTAs, however. This episode suggests that compliance with the FTA intellectual property provisions may be an issue should Congress attempt to alter U.S. law in the future.

Extent of Obligations

Most of the FTAs that the United States has joined are bilateral, although some current or contemplated FTAs extend to a limited set of trading partners. On its face, it would appear that U.S. obligations undertaken in a particular FTA would extend only to its fellow signatory or signatories. Commitments undertaken in the TRIPS Agreement may extend FTA obligations to a much broader range of U.S. trading partners, however. In particular, every WTO member may in fact be entitled to enjoy the benefits of the commitments made with a particular FTA due to the most-favored-nation obligation imposed by Article 4 of the TRIPS Agreement. That provision stipulates:

> With regard to the protection of intellectual property, any advantage, favour, privilege or immunity granted by a Member to the nationals of any other country shall be accorded immediately and unconditionally to the nationals of all other Members.

Although Article 4 of the TRIPS Agreement does provide four exceptions to this most-favored-nation obligation, none of these exceptions seems applicable to the FTAs.[128]

In view of Article 4 of the TRIPS Agreement, some experts believe that a particular FTA may extend U.S. obligations not just to its partner or partners to that agreement, but to all WTO members as well.[129] Other commentators differ, however, stating that free trade agreements are exempted from most-favored-nation obligations due to certain provisions of the General Agreement on Tariffs and Trade.[130] One possible consequence of the breadth of FTA obligations is that other trading partners may be less willing to make concessions with respect to their intellectual property laws, given that they may already be entitled to the full scope of commitments that the United States made in existing FTAs.

Fairness Concerns

Some commentators have voiced the opinion that the intellectual property provisions of the FTAs are unfair to the smaller trading partners of the United States. As explained by trade attorney George Y. Gonzalez: "The process of 'bilateral' negotiation in a 'multilateral' world context favors the negotiating stance of the affluent developed states over the wealth-constrained developing states."[131] Under this view, access to the U.S. market for agricultural products, textiles, and other basic goods is extremely important to less wealthy nations. As a result, U.S. trading partners are readily willing to accede to the intellectual property proposals of the United States during the FTA negotiation process. Some observers have criticized the United States for allegedly taking advantage of its strong bargaining position and superior intellectual property expertise when it completes FTAs.[132]

It should be appreciated, however, that many observers believe that the adoption of robust intellectual property laws lies in the best interest of U.S. trading partners, including developing nations. Commentator Jean Raymond Homere has concluded, for example, that "developing countries with stronger [intellectual property] regimes are in a better position to attract knowledge-related foreign direct investments (FDI) flows."[133] The FTAs may also encourage greater innovation, as well as the exploitation of creative works, within jurisdictions that previously lacked strong intellectual property rights.[134] It may also be appreciated that any international agreement necessarily involves an exchange of benefits and obligations among the signatory states, and that the partners to a particular FTA may be best situated to assess the advantages and disadvantages of entering into that agreement for themselves.

Concluding Observations

Congressional interest in the FTAs as a vehicle for advancing the intellectual property interests of the United States is fueled by a number of objectives. Among them are combating high levels of intellectual property piracy overseas, supporting research and development-based domestic industries, and establishing a more favorable balance of trade. Although the FTAs are believed to be an effective mechanism towards achieving these goals, concerns have arisen about both their fairness, as well as their consequences in terms of U.S. obligations towards other trading partners and their impact upon contemplated changes to domestic law. The scope of these potential consequences counsels continued congressional attention towards the FTAs that the United States has already formed, as well as those FTAs that are planned for the future.

References

[1] P.L. 107-210, 116 Stat. 993 (2002).
[2] 19 U.S.C. § 3802(b)(4)(A)(i)(II) (2004).
[3] See generally Free Trade Agreements: US Strategies and Priorities (Jeffery J. Schott, ed., 2004).

[4] See Russell J. Anderson, Jr., "Return of the Guilds: A Reflection on the Domestic and International Implications of *Eldred v. Ashcroft*," 12 *University of Baltimore Intellectual Property Law Journal* (2003), 49.
[5] See Peter K. Yu, "Currents and Crosscurrents in the International Intellectual Property Regime," 38 *Loyola of Los Angeles Law Review* (2004), 323.
[6] See Carlos M. Correa, "Investment Protection in Bilateral and Free Trade Agreements: Implications for the Granting of Compulsory Licenses," 26 *Michigan Journal of International Law* (2004), 331; CRS Report RL32375, *The U.S.-Australia Free Trade Agreement: Provisions and Implications*, by William H. Cooper.
[7] See Maria Julia Oliva, "Intellectual Property in the FTAA: Little Opportunity and Much Risk," 19 *American University International Law Review* (2003), 45.
[8] See, e.g., Frederick M. Abbott, "Intellectual Property Rights in Global Trade Framework: IP Trends in Developing Countries," 98 *American Society of International Law Proceedings* (2004), 95.
[9] See generally Roger E. Schechter and John R. Thomas, Intellectual Property: The Law of Copyrights, Patents and Trademarks (2003).
[10] *Id.*
[11] *Id.*
[12] 17 U.S.C. § 408 — 412 (2004).
[13] 17 U.S.C. § 106, 107 — 122 (2004).
[14] 17 U.S.C. § 302 (2004).
[15] 35 U.S.C. §§ 101, 102, 103 (2004).
[16] 35 U.S.C. § 101 (2004).
[17] 35 U.S.C. § 112 (2004).
[18] 35 U.S.C. § 271 (2004).
[19] 35 U.S.C. § 154(a)(2) (2004).
[20] See Restatement (Third) of Unfair Competition § 9.
[21] *Id.*
[22] *Id.* at § 18.
[23] 17 U.S.C. § 1051 (2004).
[24] See RESTATEMENT (THIRD) OF UNFAIR COMPETITION § 20.
[25] *Id.* at § 30.
[26] See CRS Report RL30989, *The U.S. Drug Approval Process: A Primer*, by Blanchard Randall IV.
[27] See G. Lee Skillington & Eric M. Solovy, "The Protection of Test and Other Data Required by Article 39.3 of the TRIPS Agreement," 24 *Northwestern Journal of International Law and Business* (2003), 1.
[28] See James T. O'Reilly, "Implications of International Drug Approval Systems on Confidentiality of Business Secrets in the U.S. Pharmaceutical Industry," 53 *Food & Drug Law Journal* (1998), 123.
[29] See Adrian Zahl, "Pharmaceuticals and the Law: As Patent Laws Converge, Attention Shifts to 'Data Protection,'" 13 *Metropolitan Corporate Counsel* no. 2 at 24 (Feb. 2005).
[30] P.L. 98-417, 98 Stat. 1585 (1984).
[31] 21 U.S.C. §355(c)(3)(E)(ii) (2004) (with respect to §505(b)(2) applications); 21 U.S.C. §355(j)(5)(F)(ii) (2004) (with respect to ANDAs).

[32] Some authorities refer to this sort of exclusivity as new molecular entity, or NME exclusivity. See Gerald J. Mossinghoff, "Overview of the Hatch-Waxman Act and Its Impact on the Drug Development Process," 54 *Food & Drug Law Journal* (1999), 187.
[33] 21 U.S.C. §355(c)(3)(E)(iii)-(iv) (2004); 21 U.S.C. §355(j)(5)(F)(iii)-(iv) (2004).
[34] P.L. 95-396, 92 Stat. 819 (1978) (codified as amended at 7 U.S.C. §§136-136y (2004)).
[35] See Aaron Xavier Fellmeth, "Secrecy, Monopoly, and Access to Pharmaceuticals in International Trade Law: Protection of Marketing Approval Data Under the TRIPS Agreement," 45 *Harvard International Law Journal* (2004), 443.
[36] See, e.g., Rebecca S. Eisenberg, "Patents and the Progress of Science: Exclusive Rights and Experimental Use," 56 *University of Chicago Law Review* (1989), 1017.
[37] See, e.g., Henry H. Perritt, Jr., "Property and Innovation in the Global Information Infrastructure," 1996 *University of Chicago Legal Forum* (1996), 261.
[38] See Frederic M. Scherer & David Ross, *Industrial Market Structure and Economic Performance* (Rand McNally & Co., 3d ed. 1990).
[39] John Perry Barlow, "The Economy of Ideas," *Wired* 2.03 (March 1994).
[40] See Congressional Research Service Report for Congress, CRS Report RL31951, *Innovation, Intellectual Property, and Industry Standards*, by John R. Thomas.
[41] See, e.g., Opinion of the Comptroller General, 159 USPQ 298, 301 (1968) ("It is a fundamental concept that territorial limitations of sovereignty preclude a country from giving extraterritorial effect to its patent laws.").
[42] See Graeme W. Austin, "Valuing 'Domestic Self-Determination' in International Intellectual Property Jurisprudence," 77 *Chicago-Kent Law Review* (2002), 1155.
[43] See Evelyn Su, *The Winners and the Losers: The Agreement on Trade — Related Aspects of Intellectual Property Rights and Its Effects on Developing Countries*, 23 HOUSTON JOURNAL OF INTERNATIONAL LAW 169 (2000).
[44] See Frederick Abbott *et al.*, *The International Intellectual Property System: Commentary and Materials* (1999).
[45] Agreement on Trade — Related Aspects of Intellectual Property Rights, Apr. 15, 1994, Annex 1C, 33 I.L.M. 1197 (1994) ["TRIPS Agreement"].
[46] P.L. 103-465, 108 Stat. 4809 (1994).
[47] TRIPS Agreement, Art. 4.
[48] See Kevin J. Nowak, "Staying Within the Negotiated Framework: Abiding by the NonDiscrimination Clause of TRIPS Article 27," 26 *Michigan Journal of International Law* (2005), 899.
[49] TRIPS Agreement, Arts. 2, 9, 35.
[50] Berne Convention for the Protection of Literary and Artistic Works, Sept. 9, 1886, 828 U.N.T.S. 221.
[51] Paris Convention for the Protection of Industrial Property, Mar. 20, 1883, 13 U.S.T. 1.
[52] May 26, 1989, I.L.M. 1477.
[53] TRIPS Agreement, Art. 12.
[54] *Id.* at Art. 10.
[55] *Id.*
[56] *Id.* at Art. 11.
[57] *Id.* at Art. 27(1).
[58] *Id.* at Art. 27(2).
[59] *Id.* at Art. 28.

[60] *Id.* at Art. 30.
[61] *Id.* at Art. 33.
[62] A "compulsory license" constitutes a statutorily created license that allows individuals to pay a royalty and use the invention without the patentee's permission. *Black's Law Dictionary* 938 (Bryan A. Gardner, ed., 8th ed. 2004).
[63] See CRS Report RS21609, *The WTO, Intellectual Property, and the Access to Medicines Controversy*, by Ian F. Fergusson.
[64] See http://www.wto.org/english/tratop_e/trips_e/implem_para6_e.htm.
[65] See also Frederick M. Abbott, "The WTO Medicines Decision: World Pharmaceutical Trade and the Protection of Public Health," 99 *American Journal of International Law* 317 (2005).
[66] See World Trade Organization, "Members OK amendment to make health flexibility permanent," (Dec. 6, 2005) (available at http://www.wto.org/english/news_e/pres05_e/pr426_e.htm).
[67] TRIPS Agreement, Art. 17.
[68] *Id.* at Art. 16.
[69] *Id.*
[70] *Id.* at Art. 18.
[71] *Id.* at Art. 21.
[72] See Lorna Brazell, *A world united? The US approach to the protection of regulatory data* (Jan. 12, 2005) (available at www.bilaterals.org).
[73] *Compare* G. Lee Skillington & Eric M. Solovy, "The Protection of Test and Other Data Required by Article 39.3 of the TRIPS Agreement," 24 *Northwestern Journal of International Law & Business* (2003), 1, *with* Carlos Maria Correa, "Unfair Competition Under the TRIPS Agreement: Protection of Data Submitted for the Registration of Pharmaceuticals," 3 *Chicago Journal of International Law* (2002), 69.
[74] TRIPS Agreement, Art. 22.1.
[75] *Black's Law Dictionary* 791 (Bryan A. Gardner, ed., 8th ed. 2004).
[76] TRIPS Agreement, Art. 39. A trade secret has been defined as a "formula, process, device, or other business information that is kept confidential to maintain an advantage over competitors." See James Pooley, *Trade Secrets* §1.01 (1998).
[77] Understanding on Rules and Procedures Governing the Settlement of Disputes, Apr. 15, 1994, WTO Agreement, Annex 2, Legal Instruments — Results of the Uruguay Round vol. 31, 33 I.L.M. 1226 (1994).
[78] Mark Clough, "The WTO Dispute Settlement System — A Practitioner's Perspective," 24 *Fordham International Law Journal* (2000), 252.
[79] TRIPS Agreement, Art. 65.
[80] See supra notes 63-65 and accompanying text.
[81] TRIPS Agreement, Art. 66.
[82] TRIPS Agreement, Art. 70.
[83] Abbott, *supra* note 8.
[84] See A. Samuel Oddi, "TRIPS — Natural Rights and a 'Polite Form of Economic Imperialism,'" 29 *Vanderbilt Journal of Transnational Law* 415 (1996).
[85] See Jean Raymond Homere, "Intellectual Property Rights Can Help Stimulate the Economic Development of Least Developed Countries," 27 *Columbia Journal of Law & the Arts* (2004), 277.

[86] See, e.g., Laurence R. Helfer, "Regime Shifting: The TRIPS Agreement and New Dynamics of International Intellectual Property Lawmaking," 29 *Yale Journal of International Law* (2004), 1.
[87] 17 U.S.C. § 302 (2004).
[88] TRIPS Agreement, Art. 12.
[89] 35 U.S.C. § 154(b) (2004).
[90] TRIPS Agreement, Art. 15.1.
[91] P.L. 107-210, 116 Stat. 993 (codified at 19 U.S.C. § 3802(b)(4)(A)(i)(II) (2004)).
[92] See Amy Kapczynski *et al.*, "Addressing Global Health Inequities: An Open Licensing Approach for University Innovations," 20 *Berkeley Technology Law Journal* (2005), 1031.
[93] See Anderson, *supra* note 4.
[94] Because the FTAs include obligations that exceed those of the TRIPS Agreement, they are commonly referred to as "TRIPS-plus agreements." See Haochen Sun, "The Road to Doha and Beyond: Some Reflections on the TRIPS Agreement and Public Health, 15 *European Journal of International Law* (2004), 146.
[95] See Frederick M. Abbott, "Toward a New Era of Objective Assessment in the Field of TRIPS and Variable Geometry for the Preservation of Multilateralism," 8 *Journal of International Economic Law* (Mar. 2005), 77.
[96] The text of each of the FTAs is available at the web site of the United States Trade Representative, www.ustr.gov.
[97] P.L. 105-304, 112 Stat. 2860 (1998).
[98] *E.g.,* U.S.-Australia FTA, Art. 17.9(9).
[99] 28 U.S.T. 7645 (June 19, 1970).
[100] U.S.-Australia FTA, Art. 17.1(2).
[101] TRIPS Agreement, Art. 33.
[102] *E.g.,* U.S.-Australia FTA, Art. 17.9(8)(a).
[103] *Id.* at Art. 17.9(7).
[104] *Id.* at Art. 17.2(2).
[105] *Id.* at Art. 17.2(10).
[106] *Id.* at Art. 17.3(1).
[107] U.S.-Chile FTA, Art. 17.10(1).
[108] See, e.g., DR-CAFTA Art. 15.10:1(c).
[109] See, e.g., U.S.-Bahrain FTA Art. 14.9(2)(a); U.S.-Jordan FTA Art. 22 n.10.
[110] U.S.-Australia FTA, Art. 17.10(2).
[111] *Id.* at Art. 17.7.
[112] U.S.-Australia FTA, Art. 17.11.
[113] See, e.g.., U.S.-Singapore FTA, Chapter 20.
[114] Yu, *supra* note 5.
[115] See Peter Drahos, "Intellectual Property and Pharmaceutical Markets: A Nodal Governance Approach," 77 *Temple Law Review* (2004), 401.
[116] J. Thomas McCarthy, "Intellectual Property — America's Overlooked Export," 20 *University of Dayton Law Review* (1995), 809.
[117] See CRS Report RL32271, *Importation of Prescription Drugs Provisions in P.L. 108-173, the Medicare Prescription Drug, Improvement, and Modernization Act of 2003*, by Susan Thaul and Donna U. Vogt.

[118] See Rahul Rajkumar, "The Central American Free Trade Agreement: An End Run Around the Doha Declaration on TRIPS and Public Health," 15 *Albany Law Journal of Science & Technology* (2005), 433.

[119] The WIPO is a specialized United Nations agency that serves as a forum for intellectual property. The WIPO website, www.wipo.int, provides more information about the agency.

[120] USPTO, "Patent Law Harmonization Talks Stall: Brazil, Argentina, India Oppose Compromise" (June 14, 2005), available at http://www.uspto.gov/main/homepagenews/bak2005jun14.htm.

[121] Drahos, *supra* note 114.

[122] Abbott, *supra* note 8.

[123] See, e.g., Warwick A. Rothnie, *Parallel Imports* (Sweet & Maxwell 1993); Simon Horner, *Parallel Imports* (Blackwell Science 1987).

[124] See Ivette P. Gomez, "Beyond the Neighborhood Drugstore: U.S. Regulation of Online Prescription Drug Sales by Foreign Businesses," 28 *Rutgers Computer & Technology Law Journal* (2002), 431.

[125] See, e.g., Seth E. Lipner, *The Legal and Economic Aspects of Gray Market Goods* (Quorum Books, Westport, Connecticut 1990).

[126] See *Fuji Photo Film Co. v. Jazz Photo Corp.*, 394 F.3d 1368 (Fed. Cir. 2005); *Jazz Photo Corp. v. United States International Trade Commission*, 264 F.3d 1094 (Fed. Cir. 2001).

[127] S. 16 at § 3(f).

[128] Article 4 of the TRIPS Agreement goes on to stipulate that:

Exempted from this obligation [of most-favored nation treatment] are any advantage, favour, privilege or immunity accorded by a Member:

(a) deriving from international agreements on judicial assistance or law enforcement of a general nature and not particularly confined to the protection of intellectual property;

(b) granted in accordance with the provisions of the Berne Convention (1971) or the Rome Convention authorizing that the treatment accorded be a function not of national treatment but of the treatment accorded in another country;

(c) in respect of the rights of performers, producers of phonograms and broadcasting organizations not provided under this Agreement;

(d) deriving from international agreements related to the protection of intellectual property which entered into force prior to the entry into force of the WTO Agreement, provided that such agreements are notified to the Council for TRIPS and do not constitute an arbitrary or unjustifiable discrimination against nationals of other Members.

As applied to the FTAs, these treaties do not appear to be general agreements concerning judicial assistance or law enforcement. They also do not appear to have been carried out under the auspices of the Berne or Rome Conventions, nor do they appear to be limited to the rights of performers, phonogram producers and broadcasting organizations. Finally, the three FTAs were completed after the WTO Agreement came into effect, rendering the final exception seemingly inapplicable.

[129] See Frederick M. Abbott, "The WTO Medicines Decision: World Pharmaceutical Trade and the Protection of Public Health," 99 *American Journal of International Law* (2005), 317.

[130] See Judy Rein, "International Governance Through Trade Agreements: Patent Protection for Essential Medicines," 21 *Northwestern Journal of International Law and Business*

[131] (2001), 379 (stating that the Article XXIV "exception provided within the GATT, permitting customs unions and free trade areas, will probably be applied to World Trade Organization agreements.").

[132] George Y. Gonzalez, "An Analysis of the Legal Implications of the Intellectual Property Provisions of the North American Free Trade Agreement," 34 *Harvard International Law Journal* (1993), 305.

[133] See Oliva, *supra* note 7.

[134] See Homere, *supra* note 85.

[135] See Keith E. Maskus, "Intellectual Property Rights and Economic Development," 32*Case Western Reserve Journal of International Law* (2000), 471.

In: New Intellectual Property Issues
Editor: John V. Luong, pp. 97-116
ISBN: 978-1-60021-211-6
© 2007 Nova Science Publishers, Inc.

Chapter 5

BIOTERRORISM COUNTERMEASURE DEVELOPMENT: ISSUES IN PATENTS AND HOMELAND SECURITY*

Wendy H. Schacht and John R. Thomas

SUMMARY

Congressional interest in the development of bioterrorism countermeasures remains strong, even after passage of legislation establishing Project BioShield. In the 109th Congress, several bills have been introduced, including S. 3, the Protecting America in the War on Terror Act, and S. 975, the Project Bioshield II Act, that would generate additional incentives for the creation of new technologies to counteract potential biological threats. These bills propose reforms to current policies and practices associated with intellectual property, particularly patents, and the marketing of pharmaceuticals and related products.

Patent ownership appears to be important in the promotion of innovation, particularly in the pharmaceutical sector. Patent title provides a limited-time monopoly over the use of a discovery in exchange for the public dissemination of information contained in the patent application. This permits the inventor to receive a return on the expenditure of resources but does not guarantee that the patent will generate commercial benefits. The requirement for patent publication is expected to stimulate additional innovation to meet similar demands in the marketplace.

Currently, the Bayh-Dole Act and the Hatch-Waxman Act include provisions that utilize patent ownership to facilitate the development and commercialization of new pharmaceuticals. The Hatch-Waxman Act also contains FDA marketing approval policies that are designed to promote the creation of new drugs. Similar market-exclusivity provisions are contained in the Orphan Drug Act.

S. 3 and S. 975 would allow for the restoration of that portion of the patent term used during the FDA approval process, and/or the extension of a patent term to reward technological innovation in the area of bioterrorism countermeasures. Pending legislation would also provide for additional FDA-administered marketing exclusivities for eligible and designated countermeasures.

Encouraging the development of new counterterrorism technologies and ensuring affordable access to new drugs and medical devices are both significant goals. These aspirations may potentially conflict, however. Introducing augmented patent- and exclusivity-based incentives may stimulate innovative firms to engage in the R&D for new

* Excerpted from CRS Report RL32917, Updated May 6, 2005

countermeasures, as well as to shepherd these products through time-consuming and costly marketing approval procedures. Commentators have expressed concern, however, over whether such heightened protections for innovators will be in proportion with the risks and costs of developing new countermeasures. Striking a balance between encouraging the development of new countermeasures and maintaining the traditional goals of our public health system is a central concern of the current discussion with respect to homeland security.

INTRODUCTION

Congressional interest in the development of bioterrorism countermeasures remains strong, even after passage of legislation establishing Project BioShield.[1] Several bills, including S. 3, the Protecting America in the War on Terror Act of 2005, and S. 975, the Project Bioshield II Act, are being considered by the 109th Congress that would generate additional incentives for the creation of new products and processes by the private sector to counteract potential biological threats. These bills propose reforms to current policies and practices associated with intellectual property, particularly patents, and the marketing of pharmaceuticals and related products.

Patents appear to be important in the promotion of innovation, particularly in the pharmaceutical sector. This report explores the role of patents in encouraging the development and commercialization of new inventions and discusses the relationships between patent ownership and the generation of biomedical products. However, the grant of a patent on a pharmaceutical does not permit marketing of the product without the approval of the Food and Drug Administration (FDA). Thus, this report also examines policies concerning the use of FDA marketing exclusivity as an additional incentive to industry research and development (R&D) in this arena. Current law and suggested legislative changes are discussed to provide a context for further exploration of the related issues.

PATENTS AND INNOVATION

Patent law is based upon the Patent Act of 1952, codified in Title 35 of the United States Code. According to the statute, one who "invents or discovers any new and useful process, machine, manufacture, or any composition of matter, or any new and useful improvement thereof, may obtain a patent therefore, subject to the conditions and requirements of this title."[2] Patents are issued by the United States Patent and Trademark Office (USPTO), generally for a term of 20 years from the date of filing. The patent grants its owner the right to exclude others from making, using, selling, offering to sell, or importing into the United States the patented invention. To be afforded patent rights, an invention must be judged to consist of patentable subject matter, possess utility, and be novel and nonobvious. The application must fully disclose and distinctly claim the invention for which protection is sought.[3]

The grant of a patent does not necessarily provide the owner with an affirmative right to market the patented invention. Pharmaceutical products are also subject to marketing approval by the Food and Drug Administration. Federal laws typically require that pharmaceutical manufacturers demonstrate that their products are safe and effective in order to bring these drugs to the marketplace. USPTO issuance of a patent and FDA marketing consent are distinct events that depend upon different criteria.

The patent system is grounded in Article I, Section 8, Clause 8 of the U.S. Constitution and is intended to stimulate new discoveries and their reduction to practice, commonly known as innovation. The Constitution states that "The Congress Shall Have Power . . . To promote the Progress of Science and useful Arts, by securing for limited Times to Authors and Inventors the exclusive Right to their respective Writings and Discoveries . . . ." The award of a patent permits the creator of an idea to exclude others, temporarily, from use of that concept without compensation. It also places the information associated with an invention within the public arena.

Patent ownership is perceived to be an incentive to innovation, the basis for the technological advancement that contributes to economic growth. It is through the commercialization and use of new products and processes that productivity gains are made and the scope and quality of goods and services are expanded. Award of a patent is intended to stimulate the investment necessary to develop an idea and bring it to the marketplace embodied in a product or process. Patent title provides the recipient with a limited-time monopoly over the use of his discovery in exchange for the public dissemination of information contained in the patent application. This is intended to permit the inventor to receive a return on the expenditure of resources leading to the discovery but does not guarantee that the patent will generate commercial benefits. The requirement for publication of the patent is expected to stimulate additional innovation and other creative means to meet similar and expanded demands in the marketplace.

Innovation produces new knowledge. One characteristic of this knowledge is that it is a "public good," a good that is not consumed when it is used. This "public good" concept underlies the U.S. patent system. As John Shoven of Stanford University points out, "the use of an idea or discovery by one person does not, in most cases, reduce the availability of that information to others."[4] Therefore the marginal social cost of the widespread application of that information is near zero because the stock of knowledge is not depleted. "Ordinarily, society maximizes its welfare through not charging for the use of a free good."[5] However, innovation typically is costly and resource intensive. Patents permit novel concepts or discoveries to become "property" when reduced to practice and therefore allow for control over their use. They ". . . create incentives that maximize the difference between the value of the intellectual property that is created and used and the social cost of its creation."[6]

Studies demonstrate that the rate of return to society as a whole generated by investments in research and development (R&D) leading to innovation is significantly larger than the benefits that can be captured by the person or organization financing the work. Some estimate that the social rate of return on R&D spending is over twice that of the rate of return to the inventor.[7] Ideas often are easily imitated as the knowledge associated with an innovation is dispersed and adapted to other products and processes that, in turn, stimulate growth in the economy. That can happen in the absence of appropriability defined as ". . . factors, excluding firm and market structure, that govern an innovator's ability to capture the profits generated by an innovation."[8] The appropriability of an invention depends on the level of competition in the industry and the type of information related to the innovation; the more competition and the more basic the knowledge, the less appropriable it is.[9] The difficulty in securing sufficient returns to spending on research and development has been associated with underinvestment in those activities.

The patent process is designed to resolve the problem of appropriability. If discoveries were universally available without a means for the inventor to realize a return on investments, there would result a ". . . much lower and indeed suboptimal level of innovation."[10] While research is often important to innovation, some commentators have noted that it can constitute less than a quarter of the cost of commercializing a new technology or technique, thus requiring the expenditure of a substantial amount of additional resources to bring most products or processes to the marketplace. The grant of a patent provides the inventor with a mechanism to capture the returns to his invention through exclusive rights on its practice for 20 years from date of filing. That is intended to allow the inventor to recoup his research spending and to encourage those investments necessary to further develop an idea and generate a marketable technology.

Issuance of a patent furnishes the inventor with a limited-time monopoly, the benefits of which are mitigated by other factors, particularly the requirements for information disclosure, the length of the patent, and the scope of rights conferred. The process of obtaining a patent places the concept on which it is based in the public domain. In return for a monopoly right to the application of the knowledge generated, the inventor must publish the ideas covered in the patent. As a disclosure system, the patent can, and often does, stimulate other firms or individuals to invent "around" existing patents to provide for parallel technical developments or meet similar market needs.

Patents may also provide a more socially desirable outcome than its chief legal alternative, trade secret protection. Trade secrecy guards against the improper appropriation of valuable, commercially useful information that is the subject of reasonable measures to preserve its secrecy.[11] Taking the steps necessary to maintain secrecy, such as implementing physical security, imposes costs that may ultimately be unproductive for society.[12] Also, while the patent law obliges inventors to disclose their inventions to the public,[13] trade secret protection requires firms to conceal them. The disclosure obligations of the patent system may better serve the objective of encouraging the diffusion of advanced technological knowledge.

The patent system thus has dual policy goals — providing incentives for inventors to invent and encouraging inventors to disclose technical information.[14] Disclosure requirements are factors in achieving a balance between current and future innovation through the patent process, as are limitations on scope, novelty mandates, and nonobviousness considerations.[15] Patents give rise to an environment of competitiveness with multiple sources of innovation, which is viewed by some experts as the basis for technological progress. This is important because, as Robert Merges (now at the University of California, Berkeley) and Richard Nelson (Columbia University) found in their studies, in a situation where only ". . . a few organizations controlled the development of a technology, technical advance appeared sluggish."[16]

Not everyone agrees that the patent system is a particularly effective means to stimulate innovation. Some observers believe that the patent system encourages industry concentration and presents a barrier to entry in some markets.[17] Others believe that the patent system too frequently attracts speculators who prefer to acquire and enforce patents rather than engage in socially productive activity.[18] Still other commentators suggest that the patent system often converts pioneering inventors into technological suppressors, who use their patents to block subsequent improvements and thereby impede technological progress.[19]

Some experts argue that patents do not work as well in reality as in theory because they do not confer perfect appropriability. In other words, they allow the inventor to obtain a larger portion of the returns on his investment but do not permit him to capture all the benefits. Patents can be circumvented and infringement cannot always be proven. Thus, patents are not the only way, nor necessarily the most efficient means, for the inventor to protect the benefits generated by his efforts. A study by Yale University's Richard Levin and his colleagues concluded that lead time, learning curve advantages (e.g. familiarity with the science and technology under consideration), and sales/service activities were typically more important in exploiting appropriability than were patents. That was true for both products and processes. However, patents were found to be better at protecting products than processes. The novel ideas associated with a product often can be determined through reverse engineering — taking the item apart to assess how it was made. That information then could be used by competitors if not covered by a patent. Because it is more difficult to identify the procedures related to a process, other means of appropriation are seen as preferable to patents, with the attendant disclosure requirements.[20]

Role of Patents in Pharmaceutical/Biomedical R&D

The utility of patents to companies varies among industrial sectors. Patents are perceived as critical in the drug industry. That may reflect the nature of R&D performed in this sector, where the resulting patents are more detailed in their claims and therefore easier to defend.[21] In contrast, one study found that in the aircraft and semiconductor industries patents are not the most successful mechanism for capturing the benefits of investments. Instead, lead time and the strength of the learning curve were determined to be more important.[22] Research undertaken by Wesley Cohen (Duke University) and his colleagues demonstrated that patents were considered the most effective method to protect inventions in the drug industry when biotechnology is included.[23]

The high cost of drug development and the concomitant uncertainty associated with clinical trials necessary for marketing approval lends significance to patents in the pharmaceutical arena. Studies by Joseph DiMasi of Tufts University and others indicate that the capitalized cost of bringing a new drug (defined as a "new molecular entity" rather than a new formulation of an existing pharmaceutical product) to the point of marketing approval is $802 million (2000 dollars).[24] Additional research done by Federal Trade Commission analysts found the costs to be even higher; between $839 million and $868 million (2000 dollars).[25] At the same time, the total capitalized costs appear to be growing at an annual rate of 7.4% above general price inflation.[26]

A large portion of new drug costs (in terms of money and time) are associated with the size and breath of clinical trials necessary to obtain FDA marketing approval. According to a study supported by the Federal Reserve of Boston, only 10% of potential drug candidates reach the human trial phase and only a small portion of these actually reach the market.[27] In research presented at a conference sponsored by the Federal Reserve Bank of Dallas, Duke University's Henry Grabowski found that only 1% of drug compounds reach the human trial stage and 22% of those entering clinical trials receive FDA approval.[28] Iain Cockburn (Boston University) notes that "as drug discovery became more science-intensive, . . .it became not just more expensive but also more difficult to manage."[29] Furthermore, returns to new drug introductions vary widely and the median new drug does not bring in sufficient profits to cover the costs of bringing the product to the marketplace.[30] According to research by Grabowski, John Vernon (Duke University), and DiMasi, only 34% of new drugs (new chemical entities) introduced generated profits that equaled the industry average R&D cost.[31]

Patents are particularly important in the pharmaceutical sector because of the relative ease of replicating the finished product. Imitation costs vary among industries. For example, while it is expensive, complicated, and time consuming to duplicate an airplane, it is relatively simple to chemically analyze a pill and reproduce it.[32] The degree to which industry perceives patents as effective has been characterized as ". . . positively correlated with the increase in duplication costs and time associated with patents."[33] In certain industries, patents significantly raise the costs incurred by nonpatent holders wishing to use the idea or invent around the patent — an estimated 40% in the pharmaceutical sector, 30% for major new chemical products, and 25% for typical chemical goods — and are thus viewed as significant. However, in other industries, patents have much smaller impact on the costs associated with imitation (e.g. in the 7%-15% range for electronics), and may be considered less successful in protecting resource investments.[34]

The significant costs of pharmaceutical R&D, coupled with the uncertainty of the clinical trial process, lend consequence to patents in this area because ". . .the disparity between the investments of innovators and those of imitators is particularly large in pharmaceuticals — almost as large as when software pirates simply copy the diskettes of an innovator."[35] While the capitalized cost of developing a new drug to the point of market approval is over $800 million, it takes only between $1 million and $2 million to obtain approval for a generic version of the pharmaceutical.[36] This difference is a result of the costs associated with clinical trials needed to demonstrate the safety and efficacy of a new drug, data that could be utilized by generic companies if not protected by a patent.[37]

LEGISLATIVE DEVELOPMENTS

In order to explore options for incentives to encourage the development and marketing of new bioterrorism countermeasures, it might be instructive to look at the state of current legislation that may be relevant to the surrounding issues. The Bayh-Dole Act and the Hatch-Waxman Act include provisions that utilize patent ownership to facilitate the development and commercialization of new pharmaceuticals (or, as is the case with the Bayh-Dole Act, all technologies). The Hatch-Waxman Act also contains FDA marketing approval policies that are designed to promote the creation of new drugs. Similar, market-exclusivity provisions are contained in the Orphan Drug Act. These laws are discussed below to provide the existing legal context within which new approaches may be considered.

The Bayh-Dole Act

P.L. 96-517, Amendments to the Patent and Trademark Act, commonly referred to as the "Bayh-Dole Act" after its two main sponsors, former Senators Birch Bayh and Robert Dole, evolved out of congressional interest in developing a uniform federal patent policy to promote the utilization of inventions made with the support of the federal research establishment.[38] Such action was deemed necessary because, at the time the legislation was under consideration, only 5% of federally-owned patents were being used. While there were possibly several reasons for such a low level of utilization (including no market applications), this was thought by many to be one consequence of the practice by most agencies of taking title to all inventions made with government funding and only permitting the nonexclusive licensing of contractor inventions.[39] Without title to inventions, or at least exclusive licenses, companies were deemed to be less likely to engage in and fund the additional R&D necessary to bring an idea to the marketplace. The Bayh-Dole Act, by providing universities, nonprofit institutions, and small businesses with ownership of patents arising from federally-funded R&D, offers an incentive for cooperative work and commercial application. Royalties derived from intellectual property rights provide the academic community an alternative way to support further research and the business sector a means to obtain a return on their financial contribution to the endeavor.[40]

Each nonprofit organization (including universities) or small business is permitted to elect (within a reasonable time frame) to retain title to any "subject invention" made as a result of R&D funded by the federal government; except under "exceptional circumstances when it is determined by the agency that restriction or elimination of the right to retain title to any subject invention will better promote the policy and objectives of this [legisltion]."[41] The owner of the intellectual property must commit to commercialization of the patent within a predetermined time frame agreed to by the supporting agency and the performing organization. As stated in the House report on one of the relevant bills, "the legislation establishes a presumption that ownership of all patent rights in government funded research will vest in any contractor who is a nonprofit research institution or a small business."[42]

Certain rights are reserved for the government to protect the public interest. The government retains ". . . a nonexclusive, nontransferable, irrevocable, paid-up license to practice or have practiced for or on behalf of the United States any subject invention throughout the world. . . ." The government also retains "march-in rights" that enable the federal agency to require the contractor (whether he owns title or has an exclusive license) to ". . . grant a nonexclusive, partially exclusive, or exclusive license in any field of use to a responsible applicant or applicants. . ." with due compensation, or to grant a license itself under certain circumstances. The special situation necessary to trigger march-in rights involves a determination that the contractor has not made efforts to commercialize within an agreed upon time frame or that the "action is necessary to alleviate health or safety needs. . ." that are not being met by the contractor.[43] To date, the government has never exercised these march-in rights.

Other provisions of the Bayh-Dole Act "authorize" the government to withhold public disclosure of information for a "reasonable time" until a patent application can be made. Licensing by any contractor retaining title under this act is restricted to companies which will manufacture substantially within the United States. Initially, universities were limited in the time they could grant exclusive licenses for patents derived from government sponsored R&D to large companies (5 of the *then* 17 years of the patent). This restriction, however, was voided by P.L. 98-620, the Trademark Clarification Act of 1984. According to Senate Report 98-662, extending the time frame for licensing to large firms ". . .is particularly important with technologies such as pharmaceuticals, where long development times and major investments are usually required prior to commercialization."[44]

The Hatch-Waxman Act

P.L. 98-417, the Drug Price Competition and Patent Term Restoration Act of 1984 (commonly known as the Hatch-Waxman Act),[45] as amended by Title XI of P.L. 108-173, the Medicare Prescription Drug and Modernization Act of 2003, made several significant changes to the patent laws as they applied to pharmaceutical products in an attempt to balance the need for innovative new drugs and the availability of less expensive generic products. The Hatch-Waxman Act established several practices intended to make it easier for generic drugs to reach the market while permitting brand name companies to recover a portion of their intellectual property rights lost during the pharmaceutical approval process.[46]

The changes legislated in the Hatch-Waxman Act include methods for extending the term of a patent to reflect regulatory delays encountered in obtaining marketing consent from the Food and Drug Administration (FDA); a statutory exemption from patent infringement for activities associated with regulatory marketing approval; mechanisms to challenge the validity of a pharmaceutical patent; and a reward for disputing the validity, enforceability, or infringement of a patent claiming an approved drug. The Hatch-Waxman Act also requires the FDA provide periods of marketing exclusivity for a pharmaceutical independent of the rights conferred by patents.

The infringement provisions in the Hatch-Waxman Act apply to pharmaceutical patents and differ from traditional procedures associated with other patented products and processes. A statutory exemption is created for certain claims of patent infringement based on acts reasonably related to seeking FDA approval to market a drug that has been patented by another firm. The company making a generic product is permitted to use data paid for and compiled by the original manufacturer to establish the drug's safety and efficacy. The generic firm must only prove that their product is bioequivalent to the innovator drug. This may allow a bioequivalent drug to reach the market as soon as the patent on the original pharmaceutical expires. Nowhere else in patent law does such a robust "experimental use" exemption exist.[47]

Additional, special provisions for addressing pharmaceutical patents are contained in the 1984 Act, including specific procedures for challenging the enforceability, validity, or infringement of approved drug patents. To encourage such patent challenges, generic applicants that file a challenge in court receive 180 days of market exclusivity provided by the FDA after that patent is found invalid, not infringed, or unenforceable or when the patent expires.

To balance such arrangements that appear to favor generic manufacturers, the Hatch-Waxman Act provides that the patent term for pharmaceuticals may be extended for a *portion* of the time lost during the FDA approval process. As noted above, ordinarily patent term is set at 20 years from the date the patent application is filed. The 1984 Act provides that for pharmaceutical patents, the patent term may be extended to reflect part of the time lost during clinical testing. More specifically, this term extension is equal to one-half the time between the effective date of the investigational new drug application and the submission of the new drug application (NDA), plus the entire time lost during FDA approval of the NDA.[48]

The Hatch-Waxman Act sets some limits on the length of the term restoration. The entire patent term restored may not exceed five years. Further, the remaining term of the restored patent following FDA approval of the NDA may not exceed 14 years.[49] The act also provides that the patentee must exercise due diligence to seek patent term restoration from the

USPTO or the period of lack of diligence will be offset from the augmented patent term.[50] Patent term extension does not occur automatically. The patent owner or its agent must file an application the USPTO requesting term extension within 60 days of obtaining FDA marketing approval.

The Hatch-Waxman Act also established so-called "marketing exclusivities" administered by the FDA. The term "marketing exclusivity" refers to a period of time during which the FDA affords an approved drug protection from competing applications for marketing approval. During this time, the FDA will not accept applications for market approval of generic versions of the brand name pharmaceutical. Two sorts of marketing exclusivities are available to innovative drug companies under the Hatch-Waxman Act: five-year new chemical entity exclusivity,[51] and three-year new clinical study exclusivity.

A drug qualifies as a new chemical entity, or NCE, if the FDA has not previously approved that drug's active ingredient.[52] The purpose of NCE exclusivity is to encourage the development of innovative drug products that include an entirely new active ingredient (commonly termed the "active moiety"), in contrast to "me-too" drugs that consist of chemical variants of previously known compounds.[53] The statute expressly stipulates that a drug does not qualify as an NCE if it consists of the salt or ester of a previously approved active ingredient as these are considered only minor chemical changes to the active ingredient.[54] NCE exclusivity prevents a subsequent generic applicant from relying upon the data submitted by the innovative drug company during a five-year period. As a result, firms are precluded from filing generic applications for five years from the date of the approval of the NDA for that active moiety.[55]

The practical effect of NCE exclusivity is to restrict a potential generic manufacturer from bringing a product to market for five years plus the length of the FDA review of the generic application.[56] If, for example, the FDA requires two years to approve a particular generic application, the real-world impact of the NCE exclusivity has been seven years of protection. In this respect NCE exclusivity operates differently from other forms of FDA-administered exclusivities, which generally prevent the FDA from approving applications, rather than accepting them in the first instance.[57]

Alternatively, a pioneer drug company may obtain a three-year new clinical study exclusivity in exchange for submitting an NDA[58] or supplemental NDA[59] that contains reports of new clinical studies of a known drug conducted by the sponsor that are essential to FDA approval of that application. The FDA has granted new clinical study exclusivity for such changes as new dosage forms, new indications, or for a switch from prescription to over-the-counter status for the drug. The purpose of the three-year new clinical study exclusivity is to encourage improvements upon drugs that are already known.

The Hatch-Waxman Act imposes four requirements that an investigation must fulfill in order to qualify for new clinical study exclusivity.[60] First, the study must be new, in that it could not have been previously used for another FDA drug approval proceeding. Second, the study must be a clinical study on humans, as compared to a preclinical study, and not a bioavailability or bioequivalence study. Third, the study must have been "conducted or sponsored" by the applicant. Finally, the study must be "essential to the approval" of the application or supplement. The FDA has defined the term "essential to the approval" as meaning "that there are no other data available that could support approval of the application."[61] A study that is interesting and provides useful background information, but not essential to approving the change in the drug, does not provide sufficient basis for an FDA award of new clinical study exclusivity.[62]

In contrast to NCE exclusivity, new clinical study exclusivity does not prevent the FDA from accepting a generic application with respect to the drug. If the new clinical study exclusivity continues to bar the issuance of marketing approval at the close of FDA review, the FDA will issue a tentative approval for the generic product that will become effective once the new clinical study exclusivity has run its course. In addition, new clinical study exclusivity only applies to the use of the product that was supported by the new clinical study.[63] If, for example, the new studies support a new indication or dosage form of the previously approved ingredient, then the three-year exclusivity applies only to that particular use or dosage form. The FDA is not barred from approving a generic drugs for other indications or dosage forms.[64]

A drug product may be subject both to NCE exclusivity and new clinical study exclusivity during the life of that product. Commonly, a new drug will initially enjoy a five-year NCE exclusivity. Later in the life of that product, the sponsor of the drug may perform additional clinical trials to qualify the drug for additional three-year exclusivities.

Orphan Drug Act

Congress enacted the Orphan Drug Act[65] in order to encourage firms to develop pharmaceuticals to treat rare diseases and conditions.[66] Such drugs are called "orphan drugs" because firms may lack the financial incentives to sponsor products to treat small patient populations.[67] Among the incentives offered via this legislation was a seven-year term of orphan drug marketing exclusivity.[68] This period commences from the date the FDA issues marketing approval on the drug.[69]

Orphan drug marketing exclusivity applies only to the indication for which the drug is approved. As a result, the FDA could approve a second application for the same drug for a different use. The FDA cannot approve the same drug made by another firm for the same use, however, unless the original sponsor approves or the original sponsor is unable to provide sufficient quantities of the drug to the market.[70]

As originally enacted, the Orphan Drug Act defined an orphan drug as one for which there was no "reasonable expectation that the cost of developing . . . will be recovered from sales in the United States of such drug."[71] Congress changed the definition to its present form in 1984.[72] Currently, in order to qualify for orphan drug status, the drug must treat a rare disease or condition (1) affecting less than 200,000 people in the United States, or (2) affecting more than 200,000 people in the United States, but for which there is no reasonable expectation that the sales of the drug would recover the costs.[73] This change allows drug sponsors to avoid showing unprofitability if the target population consisted of less than 200,000 persons.

PROPOSALS FOR CHANGE

Legislation pending before the 109th Congress would expand upon existing mechanisms in the patent and food and drug laws for encouraging the development of bioterrorism countermeasures. These bills would allow for the restoration of that portion of the patent term used during the FDA approval process, and/or the extension of a patent term to reward technological innovation in the area of bioterrorism countermeasures. These bills would also provide for additional FDA-administered marketing exclusivities for eligible and designated countermeasures.

Patent Term Fundamentals

As noted previously, several mechanisms exist for lengthening the basic 20-year patent term. Among them are Hatch-Waxman Act provisions compensating patent proprietors for delays in FDA regulatory approval proceedings.[74] This term extension is potentially available for a single patent relating to a drug product, or to a medical device, food additive, or color additive subject to regulation under the Federal Food, Drug, and Cosmetic Act.[75]

Under current law, the period of extension is ordinarily set to one-half of the clinical testing phase, less any period during which the applicant did not act with due diligence, plus the entirety of the FDA approval phase.[76] The nature of the regulated product sets the precise dates that commence the testing and approval phases that together comprise the regulatory review period. For a human drug, antibiotic, or human biological product, the clinical testing phase begins on the date the Investigational New Drug (IND) application is filed, while the FDA approval phase period starts on the date of filing of either the New Drug Application (NDA) or Product License Application (PLA).[77] For a patent claiming a medical device, the clinical testing date commences on the effective date of the Investigational Device Exemption (IDE) or, if no IDE was submitted, on the date on which the applicant began the first clinical investigation involving the device. The FDA approval phase commences on the date on which the application for product approval or notice of completion of a product development protocol under Section 515 of the Federal Food, Drug, and Cosmetic Act was initially submitted.[78] Other sorts of regulated products, such as veterinary biological products and food or color additives, are addressed in an analogous manner.

The Hatch-Waxman Act capped the maximum extension period to five years, or a total effective patent term after the extension of not more than fourteen years.[79] In addition, the Hatch-Waxman Act does not go so far as to provide a patent term extension in the usual sense — that is to say, a temporal extension of the original right to exclude others from practicing the patented invention. During the period of term extension, the rights provided by the patent are instead limited, generally speaking, to the specific use that the FDA has approved. For example, in the case of an extended product patent, the patent's rights during the extension period are generally "limited to any use approved for the product" that subjected it to regulatory approval delays at the FDA.[80]

Proposed Patent Term Restoration Reforms

Pending legislation includes provisions that act similarly with respect to patent term restoration on bioterrorism countermeasures, although they potentially allow for longer periods of expansion of patent term than is currently available under the Hatch-Waxman Act.

It should be noted that the Hatch-Waxman Act refers to the lengthening of patent life in compensation for delays in FDA marketing approval as an "extension of patent term."[81] However, the pending legislation uses the word "restoration" to refer to this concept, while employing the word "extension" to refer to a distinct reward for technological innovation in the area of bioterrorism countermeasures.

Both S. 3, The Protecting America in the War on Terror Act of 2005, and S. 975, The Project Bioshield II Act, would introduce a new 35 U.S.C. § 156a into the Patent Act of 1952. As with current law, these statutes would lengthen an eligible patent on a bioterrorism countermeasure on a day-per-day basis for the time lost during the FDA approval phase. These bills would also extend patent term on a day-per-day basis during the entire period of the clinical testing phase, however, rather than merely for one-half of that period as provided in the Hatch-Waxman Act.[82] In contrast to the Hatch-Waxman Act, the pending legislation also places no cap upon the maximum period of increase in patent term.

Notably, the Hatch-Waxman Act sets the first day of the regulatory review period as the date that the drug sponsor's commences clinical studies at the FDA. However, both S. 3 and S. 975 consider the regulatory review period to commence on the later of either the date the clinical testing phase begins at the FDA, or the date upon which the USPTO issues the patent sought to be extended.[83] In circumstances where the USPTO issues the patent after the sponsor has commenced clinical studies at the FDA, the relevant regulatory review period under current law may actually be longer than under the proposed legislation.

Both S. 3 and S. 975 stipulate that this patent term restoration is mutually exclusive with respect to "extension of patent term" available under current law.[84] As a result, only a single patent that claims a countermeasure product may be lengthened by either the period established by the "extension of patent term" provision of the existing Hatch-Waxman Act or the new "restoration of patent terms relating to countermeasure products" under proposed 35 U.S.C. § 156a, but not both.

Proposed Patent Term Extension Reforms

Pending legislation would also create the possibility of "patent term extension," codified at 35 U.S.C. § 158, that has no direct analogy under current law. Under S. 3 and S. 975, this term extension is available to entities that have developed FDA-approved countermeasure products. The period of term extension under both bills ranges from a minimum of six months to a maximum of two years.[85] The Secretary of Health and Human Services would be granted discretion to set the period of term extension based upon such factors as the nature of the threat to be countered, the difficulty and expense in developing the countermeasure, and the impact of patent extension upon consumers and healthcare providers.[86]

Considerable discussion has occurred with respect to whether the term extension provided by proposed 35 U.S.C. § 158 must be applied to a patent claiming the countermeasure itself, or instead to any one patent that the countermeasure innovator selects within its intellectual property portfolio. The concept that a countermeasure innovator could enjoy expanded patent life with respect to an unrelated product has been termed a "wild card" term extension.[87] Under this regime, a countermeasure innovator could potentially obtain a term extension of a patent relating to a bestselling mainstream pharmaceutical, thereby shielding it from generic competition for an additional period of six months to two years.

Supporters of a wild card patent term extension urge that patent-based incentives on profitable products, such as "blockbuster" drugs, are needed to encourage firms to develop potentially less profitable countermeasures.[88] Detractors have expressed concerns over the equities of requiring patients to finance the development of countermeasures through the purchase of medications to treat their particular illnesses or medical conditions.[89]

S. 975 appears to employ this "wild card" patent term extension concept. That bill stipulates that patents eligible for the extension period of six months of two years must relate to "designated products," with that term in turn generally defined as any drug, antibiotic drug, device, or biological product.[90] This definition contrasts with that provided with respect to term restoration as compensation for FDA regulatory review, which expressly states that the patent must claim a countermeasure product.[91]

The situation with respect to S. 3 is less clear. In a statement on the floor of the Senate, Senator Judd Gregg appears to disavow the notion that S. 3 creates "wild card" exclusivity. According to that statement, S. 3 creates "additional incentives involving marketing exclusivity that could be granted for up to two years for the product that was used as a countermeasure. This is an important distinction from the so-called 'wild card' exclusivity idea, which would allow a company to extend the patent protection of a different product ...." [92] It should be noted, however, that S. 3 expressly restricts the availability of term restoration to patents that claim countermeasure products, but incorporates no similar restriction with regard to term extension.[93] As a result, some observers have expressed concerned about whether the bill provides "wild card" exclusivity or not.[94]

Proposed Marketing Exclusivity Reforms

As noted previously, the food and drug laws provide for certain FDA-administered marketing exclusivities. Among these exclusivities is the five-year new chemical entity exclusivity,[95] a three-year new clinical trials exclusivity,[96] and a seven-year orphan drug exclusivity.[97] Subject to certain exceptions, these exclusivities generally prevent the FDA from granting marketing approval with respect to another sponsor's competing drug during the statutory period.[98]

S. 975 would expand upon these existing mechanisms with respect to new drugs that were developed by certified researchers and that qualify as countermeasure products. In particular, the five-year new chemical entity exclusivity would be doubled to ten years; the three-year new clinical studies exclusivity would be doubled to six years; and the seven-year orphan drug exclusivity would be extended to ten years.[99] No comparable provision appears in S. 3.

March-In Rights

As discussed above, the Bayh-Dole Act provides the government with the right to "march-in" and license to another manufacturer a patent that was originally developed by a contractor during federally-funded R&D under certain, very specific conditions. This right has never been exercised. However, some in industry view this provision as a potential barrier to on-going innovation. S. 975 would permit the owner of a patent related to a countermeasure made under the conditions triggered by the Bayh-Dole Act to request that the government agency that funded the original research waive its march-in rights.[100] No comparable provision appears in S. 3.

It should be noted that in addition to the march-in rights afforded to the government under the Bayh-Dole Act, the government also has the authority to take private property for public use under "eminent domain."[101] While most frequently applied to real estate, the general principle generally applies to intellectual property as well.[102] As a result, the U.S. government effectively enjoys the ability to declare a "compulsory license" that allows it to use a patented invention without obtaining the permission of the patentee. In turn, the federal government has consented to a suit by private patent owners in order to obtain compensation for government uses.[103] Section 1498(a) of Title 28 of the U.S. Code provides in part:

> Whenever an invention described in and covered by a patent of the United States is used or manufactured by or for the United States without license of the owner thereof or lawful right to use or manufacture the same, the owner's remedy shall be by action against the United States in the United States Claims Court for the recovery of his reasonable and entire compensation for such use and manufacture.

The remaining paragraphs of § 1498 provide analogous provisions pertaining to other intellectual property rights, including copyright, plant variety protection certificates, and semiconductor mark works.[104]

Under § 1498(a), all patent suits against the U.S. government are litigated in the U.S. Court of Federal Claims. These Court proceedings are conducted using the same general standards as does litigation between private parties. The patent owner represents itself, while the Attorney General and Department of Justice are responsible for representing the U.S. government in § 1498 cases.[105] Unlike private patent suits, however, there are no jury trials in § 1498 cases.[106] Appeals from the United States Court of Claims proceed to the U.S. Court of Appeals for the Federal Circuit.[107]

As compared to remedies available in patent infringement suits against private parties, the remedies available in § 1498(a) suits are more limited. In private patent litigation, the adjudicated infringer is ordinarily enjoined from using the patented invention throughout the remaining term of the patent.[108] The adjudicated infringer may also have to compensate the patent owner for profits lost due to the infringement.[109] Additionally, if a court deems the defendant to have been a "willful infringer," the court may order the defendant to pay the patent owner up to three times the actual damages suffered.[110]

In contrast, § 1498(a) limits available remedies to "reasonable and entire compensation" to the patent owner. As a result, the government may not be enjoined from practicing a patented invention. The courts have also generally limited the damages that the government must pay to the patentee to the level of a "reasonable royalty."[111] A "reasonable royalty" for purposes of patent infringement damages is "the amount that a person desiring to manufacture or use a patented article, as a business proposition, would be willing to pay as a royalty and yet be able to make or use the patented article, in the market at a reasonable profit."[112] Finally, tripled damages for willful infringement are not available against the government.[113]

Concluding Observations

The use of patent ownership and marketing exclusivity to encourage innovation in the pharmaceutical industry is reflected in existing law and government policy. Studies have shown that these efforts appear to have been successful in facilitating the development and commercialization of new technologies.[114] As such, there appears to be congressional interest in utilizing and expanding this approach for the specific purpose of generating innovative bioterrorism countermeasures. Two bills introduced in the 109th Congress, S. 3 and S. 975, establish regimes under which patent terms are restored and/or extended and market exclusivity periods increased to reward innovation in this arena. Although different in the specific provisions, these bills build upon incentives that have been used in the past.

Encouraging the development of new counterterrorism technologies, on one hand, and ensuring affordable access to new drugs and medical devices, on the other, are both significant goals. These aspirations may potentially conflict, however. Introducing augmented patent- and exclusivity-based incentives may stimulate innovative firms to engage in the research and development of new countermeasures, as well as to shepherd these products through time-consuming and costly marketing approval procedures. Commentators have expressed concern, however, over whether such heightened protections for innovators will be in proportion with the risks and costs of developing new countermeasures. Detractors have also questioned the propriety of financing the development of antiterrorism technologies through increased prices upon distinct medical products via the wild card patent term extension. Striking a balance between encouraging the development of new countermeasures and maintaining the traditional goals of our public health system is a central concern of the current discussion with respect to homeland security.

References

[1] P.L. 108-276, see also CRS Report RS21507, *Project BioShield*, by Frank Gottron.

[2] 35 U.S.C.§101

[3] For more information see CRS Report RL30756, *Patent Law and Its Application to the Pharmaceutical Industry: An Examination of the Drug Price Competition and Patent Term Restoration Act of 1984 ("The Hatch-Waxman Act")*, by Wendy H. Schacht and John R. Thomas.

[4] John B. Shoven, "Intellectual Property Rights and Economic Growth," in *Intellectual Property Rights and Capital Formation in the Next Decade*, eds. Charls E. Walker and Mark A. Bloomfield (New York: University Press of America, 1988), 46.

[5] Robert P. Benko, "Intellectual Property Rights and New Technologies," in *Intellectual Property Rights and Capital Formation in the Next Decade*, 27.

[6] Stanley M. Besen and Leo J. Raskind, "An Introduction to the Law and Economics of Intellectual Property," *Journal of Economic Perspectives*, Winter 1991, 5.

[7] For a list of relevant research in this area see Council of Economic Advisors. *Supporting Research and Development to Promote Economic Growth: The Federal Government's Role*, October 1995, 6-7.

[8] David J. Teece, "Profiting from Technological Innovation: Implications for Integration, Collaboration, Licensing, and Public Policy," in *The Competitive Challenge,* ed. David J. Teece, (Cambridge: Ballinger Publishing Co., 1987), 188.
[9] Edwin Mansfield. "Intellectual Property Rights, Technological Change, and Economic Growth," in, *Intellectual Property Rights and Capital Formation in the Next Decade,* 10-11.
[10] Kenneth W. Dam, "The Economic Underpinnings of Patent Law," *Journal of Legal Studies*, January 1994, 247.
[11] American Law Institute, Restatement of Unfair Competition Third §39, 1995.
[12] David D. Friedman, et al., "Some Economics of Trade Secret Law," 5 *Journal of Economic Perspectives*, 1991, 61.
[13] 35 U.S.C. §112 (2000).
[14] Robert P. Merges, "Commercial Success and Patent Standards: Economic Perspectives on Innovation," *California Law Review*, July 1988, 876.
[15] *The Economic Underpinnings of Patent Law*, 266-267. Scope is determined by the number of claims made in a patent. Claims are the technical descriptions associated with the invention. In order for an idea to receive a patent, the law requires that it be ". . .new, useful [novel], and nonobvious to a person of ordinary skill in the art to which the invention pertains."
[16] Robert P. Merges and Richard R. Nelson, "On the Complex Economics of Patent Scope," *Columbia Law Review*, May 1990, 908.
[17] See John R. Thomas, "Collusion and Collective Action in the Patent System: A Proposal for Patent Bounties," *University of Illinois Law Review*, 2001, 305.
[18] Ibid.
[19] *On the Complex Economics of Patent Scope*, 839.
[20] Richard C. Levin, Alvin K. Klevorick, Richard R. Nelson, and Sidney G. Winter. "Appropriating the Returns for Industrial Research and Development," *Brookings Papers on Economic Activity*, 1987, in *The Economics of Technical Change*, eds. Edwin Mansfield and Elizabeth Mansfield (Vermont, Edward Elgar Publishing Co., 1993), 254.
[21] Ibid., 255 and 257. See also: *Intellectual Property Rights, Technological Change, and Economic Growth,* 12 and 13.
[22] *Appropriating the Returns for Industrial Research and Development*, 253.
[23] Wesley M. Cohen, Richard R. Nelson, and John P. Walsh, *Protecting Their Intellectual Assets: Appropriability Conditions and Why U.S. Manufacturing Firms Patent (or Not)*, NBER Working Paper 7552, Cambridge, National Bureau of Economic Research, February 2000, available at [http://www.nber.org/papers/w7552].
[24] Joseph A. DiMasi, Ronald W. Hansen, and Henry G. Grabowski. "The Price of Innovation: New Estimates of Drug Development Costs," 22 *Journal of Health Economics,* 2003. Capitalized cost includes the "time cost" associated with an investment and the cost of testing drug products that fail.
[25] Christopher P. Adams and Van V. Brantner, *Estimating the Costs of New Drug Development: Is it Really $802m?*, Federal Trade Commission, December 2004, available at [http://media.romanvenable.net/images/drugCost.pdf].
[26] *The Price of Innovation: New Estimates of Drug Development Costs*, 180

[27] Carrie Conway, "The Pros and Cons of Pharmaceutical Patents," *Regional Review*, Federal Reserve Bank of Boston, March, 2003, available at [http://www.findarticles.com].
[28] Henry G. Grabowski, "Patents, Innovation, and Access to New Pharmaceuticals," *Journal of International Economic Law,* 2002, 851.
[29] Iain Cockburn, "The Changing Structure of the Pharmaceutical Industry," *Health Affairs*, Janurary/February 2004, 15.
[30] Henry G. Grabowski, "Patents and New Product Development in the Pharmaceutical and Biotechnology Industries," *Science and Cents: Exploring the Economics of Biotechnology*, Proceedings of a 2002 Conference, Federal Reserve Bank of Dallas, 95-96 available at [http://www.dallasfed.org/research/pubs/science/grabowski.pdf] and Henry Grabowski, John Vernon, and Joseph A. DiMasi, "Returns on Research and Development for 1990s New Drug Introductions," 20 *Pharmacoeconomics*, 2002.
[31] *Returns on Research and Development for 1990s New Drug Introductions*, 23.
[32] Federic M. Scherer, "The Economics of Human Gene Patents", 77 *Academic Medicine,* December 2002, 1350.
[33] *Appropriating the Returns for Industrial Research and Development.*, 269.
[34] Edwin Mansfield, Mark Schwartz, and Samuel Wagner. "Imitation Costs and Patents: An Empirical Study," *The Economic Journal*, December 1981, in *The Economics of Technical Change*, 270.
[35] *The Economics of Human Gene Patents*, 1352.
[36] *Patents, Innovation, and Access to New Pharmaceuticals*, 852.
[37] *The Economics of Human Gene Patents*, 1352
[38] House Committee on Science and Technology, *Government Patent Policy*, 95th Cong., 2nd sess., May 1978, H.Rept. Prt. 4.
[39] *Ibid.*, 5.
[40] For discussion of this law, additional provisions, and implementation see CRS Report RL32076, *The Bayh-Dole Act: Selected Issues in Patent Policy and the Commercialization of Technology*, by Wendy H. Schacht and CRS Report RL32324, *Federal R&D, Drug Discovery, and Pricing: Insights from the NIH-University-Industry Relationship*, by Wendy H. Schacht.
[41] *Government Patent Policy*, 5.
[42] House Committee on the Judiciary, *Report to Accompany H.R. 6933*, 96th Congress., 2d. Sess., H.Rept. 96-1307, Part 1, 3.
[43] 15 U.S.C. §203.
[44] Senate Committee on the Judiciary, *Report to Accompany S. 2171*, 98th Cong., 2d Sess. S.Rept. 98-662, 1984, 3.
[45] 21 U.S.C. § 355 and following.
[46] For a detailed discussion of the Hatch-Waxman Act see CRS Report RL30756, *Patent Law and Its Application to the Pharmaceutical Industry: An Examination of the Drug Price Competition and Patent Term Restoration Act of 1984 (The Hatch-Waxman Act)*, by Wendy H. Schacht and John R. Thomas, and CRS Report RL32377, *The Hatch-Waxman Act: Legislative Changes Affecting Pharmaceutical Patents*, by Wendy H. Schacht and John R. Thomas.
[47] See CRS Report RL32651, *Scientific Research and the Experimental Use Privilege in Patent Law*, by John R. Thomas.

[48] 25 U.S.C. § 156.
[49] 35 U.S.C. § 156(c).
[50] 35 U.S.C. § 156(d)(2)(B)
[51] Some authorities refer to this sort of exclusivity as new molecular entity, or NME exclusivity. See Gerald J. Mossinghoff, "Overview of the Hatch-Waxman Act and Its Impact on the Drug Development Process," 54 *Food and Drug Law Journal,* 1999, 187.
[52] See 21 C.F.R. §314.108(a).
[53] See *Abbott Labs. v. Young*, 920 F.2d 984, 986 (D.C. Cir. 1990)
[54] 21 U.S.C. §3 55(c)(3)(E) (2004) (with respect to §505(b)(2) applications); 21 U.S.C. § 355(j)(5)(F) (2004) (with respect to ANDAs).
[55] 21 U.S.C. § 355(c)(3)(E)(ii) (with respect to §505(b)(2) applications); 21 U.S.C. § 355(j)(5)(F)(ii) (with respect to ANDAs).
[56] See Marvin M. Goldenberg, "Medicare and the New Generic Drug Legislation," 29 *Pharmacy and Therapeutics* no. 2 at 89,February 2004, 90.
[57] See Elizabeth H. Dickinson, "FDA's Role in Making Exclusivity Determinations," 54 *Food & Drug Law Journal,* 1999, 195.
[58] 21 U.S.C. §355(c)(3)(E)(iii) (with respect to §505(b)(2) applications); 21 U.S.C. § 355(j)(5)(F)(iii) (with respect to ANDAs).
[59] 21 U.S.C. § 355(c)(3)(E)(iv) (with respect to §505(b)(2) applications); 21 U.S.C. § 355(j)(5)(F)(iv) (with respect to ANDAs).
[60] 21 U.S.C. § 355(c)(3)(E)(iii)-(iv) (with respect to §505(b)(2) applications); 21 U.S.C. § 355(j)(5)(F)(iii)-(iv) (with respect to ANDAs).
[61] See 21 C.F.R. § 314.108(a)
[62] See Upjohn Co. v. Kessler, 938 F. Supp. 439 (W.D. Mich. 1996).
[63] See Valerie Junod, "Drug Marketing Exclusivity Under United States and European Union Law," 59 *Food and Drug Law Journal,* 2004, 479.
[64] See Thomas J. Parker *et al.*, "FDA Marketing Exclusivity for Single Enantiomers of Previously Approved Racemates," 15 *Journal of Proprietary Rights* no. 1 at 8, January 2003.
[65] Orphan Drug Act, Pub. L. No. 97-414, 96 Stat. 2049 (1982) (codified as amended at 21 U.S.C. §§ 360aa-360ee, 26 U.S.C. § 45C, 42 U.S.C. § 236).
[66] See, e.g., Gary A. Pulsinelli, "The Orphan Drug Act: What's Right With It," 15 *Santa Clara Computer & High Technology Law Journal,* 1999, 299. The Orphan Drug Act also applies to varying degrees to biologics and medical devices, which, along with drugs, are commonly termed "orphan products."
[67] *See, e.g.,* David B. Clissold, "Prescription for the Orphan Drug Act: The Impact of the FDA's 1992 Regulations and the Latest Congressional Proposals for Reform," 50 *Food & Drug Law Journal,* 1995, 125.
[68] 21 U.S.C. §360cc.
[69] *Id.*
[70] 21 U.S.C. § 360cc(b).
[71] Orphan Drug Act, Pub. L. No. 97-414, § 526(a)(2), 96 Stat. 2049 (1982) (codified as amended at 21 U.S.C. § 360bb(a)(2) (2000)).
[72] Health Promotion and Disease Prevention Amendments of 1984, P.L. 98-551, 98 Stat. 2815 (1984).
[73] 21 U.S.C. § 360bb(a)(2).

[74] 35 U.S.C. § 156 (2004).
[75] 35 U.S.C. § 156(f) (2004).
[76] 35 U.S.C. § 156(c) (2004).
[77] 37 C.F.R. § 1.740(a)(10)(I) (2004).
[78] 37 C.F.R. § 1.740(a)(10)(v) (2004).
[79] 35 U.S.C. § 156(b) (2004).
[80] 35 U.S.C. § 156(b)(1) (2004).
[81] 35 U.S.C. § 156 (2004).
[82] S. 3 at § 113(c) (introducing 35 U.S.C. § 156a(b)); S. 975 at § 331(b) (introducing 35 U.S.C. § 156a(b))).
[83] S. 3 at § 113(c) (introducing 35 U.S.C. § 156a(a)(2)); S. 975 at § 331(b) (introducing 35 U.S.C. § 156a(a)(2)).
[84] S. 3 at § 113(c) (introducing 35 U.S.C. § 156a(b)(4)); S. 975 at § 331(b) (introducing 35 U.S.C. § 156a(b)(4)).
[85] S. 3 at § 113(c) (introducing 35 U.S.C. § 158(a)(d)(1)); S. 975 at § 301(b)(4)(a)(iv).
[86] S. 3 at § 113(d) (introducing 35 U.S.C. § 158(a)(d)(2)); S. 975 at § 301(b)(4)(a)(ii).
[87] See Chris Mondics, "Drug firms on Senate agenda," *Philadelphia Inquirer,* April 10, 2005, A5.
[88] McKenna, Long & Aldridge, Press Release, *'Bioshield' Drug-Patent Plan Draws Fire,* April 1, 2005 available at [http://www.mckennalong.com/news-pressrelease-1235.html].
[89] Generic Pharmaceutical Association, Press Release, *Legislation Promoted as a Countermeasure Against Bioterrorism Would Counter Bipartisan Measures to Constrain Prescription Costs,* February 8, 2005 available at [http://www.gphaonline.org/].
[90] S. 975 at § 331(c) (introducing 35 U.S.C. § 158(a)(3)).
[91] *Id.* at § 331(b) (introducing 35 U.S.C. § 156a(a)(3)(A)).
[92] Senator Judd Gregg, *Congressional Record,* April 6, 2005, S3264.
[93] Compare S. 3, § 113(c) (introducing 35 U.S.C. § 156a(a)(3)) with S. 3, § 113(d) (introducing 35 U.S.C. § 158(a)(2)).
[94] See "GphA to Gregg: Clarify That 'Wildcard Exclusivity' Not in S. 3," *INSIDE CMS,* March 10, 2005.
[95] See 21 U.S.C. § 355j(c)(3)(E)(ii) (2004) (with respect to § 505(b)(2) applications); 21 U.S.C. § 355(j)(5)(F)(ii) (2004) (with respect to Abbreviated New Drug Applications).
[96] See 21 U.S.C. § 355j(c)(3)(E)(iii), (iv) (2004) (with respect to § 505(b)(2) applications); 21 U.S.C. § 355(j)(5)(F)(iii), (iv) (2004) (with respect to Abbreviated New Drug Applications).
[97] See 21 U.S.C. § 360cc(a) (2004).
[98] In addition, under the five-year new chemical entity exclusivity, the FDA will not consider applications for a generic version of a new chemical entity for five years after approval of the original. As a result, the practical period of exclusivity is five years plus the time that the FDA takes to review the competitor's application. See Valerie Junod, *Drug Marketing Exclusivity Under United States and European Union Law*, 59 FOOD & DRUG L.J., 2004, 479.
[99] S. 975 at § 331(e) (introducing 21 U.S.C. § 505C).
[100] S. 975 at 331(d)

[101] For a discussion of this and related topis see CRS Report RL32051, *Innovation and Intellectual Property Issues in Homeland Security*, by John R. Thomas.

[102] Thomas F. Cotter, "Do Federal Uses of Intellectual Property Implicate the Fifth Amendment?," 50 *Florida Law Review,* 1998, 529.

[103] See Lionel Marks Lavenue, "Patent Infringement Against the United States and Government Contractors Under 28 U.S.C. § 1498(a) in the United States Court of Federal Claims," 2 *Journal of Intellectual Property Law,* 1995, 389.

[104] 28 U.S.C. § 1498 (2000).

[105] 28 U.S.C. § 516 (2000).

[106] 28 U.S.C. § 2402 (2000).

[107] 28 U.S.C. § 1295(a)(3) (2000).

[108] 28 U.S.C. § 283 (2000).

[109] See *Panduit Corp. v. Stahlin Bros. Fibre Works, Inc.*, 575 F.2d 1152 (6th Cir. 1978).

[110] 35 U.S.C. § 284 (2000).

[111] See Tektronix, Inc. v. United States, 552 F.2d 343 (Ct. Cl. 1977). Some more recent precedent has suggested that in some cases, the U.S. government may be obliged to pay the full lost profits of the patentee rather than a reasonable royalty. See Gargoyles, Inc. v. United States, 113 F.3d 1572 (Fed. Cir. 1997). However, reportedly the last instance that an award of lost profits was made for government use of a patented invention was in 1930. David M. Schlitz & Richard J. McGrath, "Patent Infringement Claims Against the United States Government," 9 *Federal Circuit Bar Journal,* 2000, 351.

[112] Wright v. United States, 53 Fed. Cl. 466 (2002).

[113] DeGraffenried v. United States, 228 Ct. Cl. 780 (1981).

[114] See *The Bayh-Dole Act: Selected Issues in Patent Policy and the Commercialization of Technology* and *Patent Law and Its Application to the Pharmaceutical Industry: An Examination of the Drug Price Competition and Patent Term Restoration Act of 1984 (The Hatch-Waxman Act).*

In: New Intellectual Property Issues
Editor: John V. Luong, pp. 117-127
ISBN: 978-1-60021-211-6
© 2007 Nova Science Publishers, Inc.

Chapter 6

SAFE HARBOR FOR PRECLINICAL USE OF PATENTED INVENTIONS IN DRUG RESEARCH AND DEVELOPMENT: *MERCK KGAA V. INTEGRA LIFESCIENCES I, LTD.*[*]

Brian T. Yeh

SUMMARY

In *Merck KGaA v. Integra Lifesciences I, Ltd.*, __ U.S. __, 125 S. Ct. 2372 (2005), the United States Supreme Court unanimously held that the preclinical use of patented inventions in drug research is exempted from patent infringement claims by the "safe harbor" provision of the Patent Act, 35 U.S.C. § 271(e)(1). (Merck KGaA is a German company unaffiliated with the U.S.-based pharmaceutical company Merck & Co.) This decision potentially may help expedite the development of new medical treatments and lower the cost of some drugs for consumers.

In 2003, the U.S. Court of Appeals for the Federal Circuit had narrowly construed the safe harbor provision as protecting only clinical research activities that produce information for submission to the Food and Drug Administration (FDA) in the regulatory process. In vacating that decision, the U.S. Supreme Court ruled that the exemption applies to *all* uses of patented inventions that are "reasonably related" to the process of developing any information for FDA submission. The Court explained that, under certain conditions, the safe harbor provision is even "sufficiently broad" to protect the use of patented compounds in experiments that are not ultimately submitted to the FDA or drug experiments that are not ultimately the subject of an FDA submission. Finally, the scope of the exemption is not limited only to preclinical studies pertaining to a drug's safety in humans, but also includes preclinical data regarding a drug's efficacy, mechanism of action, pharmacokinetics, and pharmacology.

However, the Court cautioned that the exemption does not reach all experimental activity that at some point, however attenuated, may lead to an FDA approval process. For example, the safe harbor provision does not embrace basic scientific research performed on a patented compound without the intent to develop a particular drug or without a reasonable belief that the compound will cause a particular physiological effect that the researcher desires. In

[*] Excerpted from CRS Report RL33114, Updated October 7, 2005

addition, because the matter was not at issue in the case, the Court expressly declined to decide whether or to what extent the exemption applies to patented "research tools" that are often used to facilitate general research in developing compounds for FDA submissions.

INTRODUCTION

In *Merck KGaA v. Integra Lifesciences I, Ltd.*, __ U.S. __, 125 S. Ct. 2372 (2005), the United States Supreme Court decided, without dissent, that the patent law's safe harbor provision exempts from infringement the preclinical use of patented inventions in drug research. Without this legal immunity, pharmaceutical companies face patent infringement liability when they conduct preclinical experiments using rival companies' patented compounds. The U.S. Court of Appeals for the Federal Circuit had earlier found that the statutory exemption applied only to clinical research activity that contributes "relatively directly" to information the Food and Drug Administration (FDA) considers in approving a drug.[1] This narrow interpretation of the safe harbor provision had raised concerns that the patent law could significantly restrict the development and introduction of new medical treatments and generic drugs.

Vacating the appellate court's decision, the U.S. Supreme Court unanimously ruled that the exemption protects all uses of patented inventions that are "reasonably related" to the process of developing any information for FDA submission, which includes preclinical studies. The Court's expansive construction of the safe harbor provision "leaves adequate space for experimentation and failure on the road to regulatory approval"[2] and "provides a wide berth for the use of patented drugs in activities related to the federal regulatory process."[3]

BACKGROUND

It is normally a violation of the Patent Act to use any patented invention without prior authorization of the patent owner.[4] However, a statutory exception to this general rule provides: "It shall not be an act of infringement to make, use, offer to sell, or sell within the United States or import into the United States a patented invention ... solely for uses reasonably related to the development and submission of information" to the United States Food and Drug Administration (FDA).[5] Thus, a party that uses a patented invention without the patent owner's permission is committing an infringing act, but if the use comes within the scope of the statutory exception, the party will not be held liable for violating the patent owner's rights.

The Hatch-Waxman Act. The statutory exception was created by the Drug Price Competition and Patent Term Restoration Act of 1984,[6] commonly known as the Hatch-Waxman Act. This legislation modified the Patent Act by creating a new section, 35 U.S.C. § 271(e), that provides "safe harbor" from infringement for pharmaceutical companies using patented inventions in their drug research and development operations.

The Hatch-Waxman Act is widely credited with encouraging and expediting the creation and availability of generic versions of approved patented drugs. Prior to its enactment, pharmaceutical companies had to wait until all relevant patents expired before undertaking the clinical research necessary to obtain FDA approval of generic equivalents.[7] Thus, an established drug's patent term was *de facto* extended beyond its expiration date by the length of the FDA regulatory process for approving the generic equivalent, which took more than two years.[8] The Hatch-Waxman Act allows generic drug manufacturers to conduct safety and effectiveness tests during the time the brand name drug's patent is still in force, often resulting in immediate introduction of a generic drug into the market upon the pioneer drug's patent expiration.[9]

The FDA Drug Approval Process. The Federal Food, Drug, and Cosmetics Act (FDCA) regulates the manufacture, use, or sale of drugs.[10] Under the FDCA, the FDA must determine that a drug is safe and effective before it can be marketed to consumers. The FDCA establishes a two-stage approval process for new drugs: an "Investigational New Drug" (IND) application and a "New Drug Application" (NDA).[11]

The drug manufacturer must file an IND with the FDA after the company has identified, through preclinical testing on animals and in test tubes, chemical compounds that appear to have beneficial therapeutic effects. The IND is a request for authorization to conduct clinical (human) testing, and it must contain information and data from the preclinical studies that justify the proposed clinical trial.[12]

Once the FDA approves the IND, the drugmaker can commence clinical studies. If these studies demonstrate that a new drug is reasonably safe and effective for use, the drugmaker is required to submit a NDA.[13] The NDA must include data from preclinical and clinical studies. After extensive review of the NDA, the FDA issues final approval or denial of the application for manufacturing and selling the new drug to the public.[14]

The Scope of Safe Harbor

The Patent Act's safe harbor provision has often been compared to the "fair use" defense in copyright law, since it immunizes from liability otherwise infringing acts in order to advance compelling public policy interests. The legislative history of the Hatch-Waxman Act provides the basis for this analogy: "Just as we have recognized the doctrine of fair use in copyright, it is appropriate to create a similar mechanism in the patent law. That is all this bill does."[15] Despite this deceptively simple language of purpose, the safe harbor provision has been the subject of confusion and litigation for many years following its enactment. For over two decades, federal courts struggled to define the breadth and contours of the exemption, particularly concerning the types and uses of patented invention covered by the safe harbor.

Types Covered. As for the types of covered patented invention, the United States Supreme Court in *Eli Lilly & Co. v. Medtronic, Inc.* expansively interpreted § 271(e)(1) to include not only drug and veterinary products, but also medical devices that are subject to pre-market approval by the FDA.[16] The *Eli Lilly* Court determined that "[t]he phrase 'patented invention' in § 271(e)(1) is defined to include all inventions, not drug-related inventions alone."[17] The Court opined that if Congress had wanted the safe harbor to cover only generic drugs, "there were available such infinitely more clear and simply ways of expressing that intent."[18] As written, § 271(e)(1) applies to the "entire statutory scheme of regulation,"[19] including "medical devices, food additives, color additives, new drugs, antibiotic drugs, and human biological products."[20]

Uses Covered. Concerning the protected uses of a patented invention, a long disputed issue was what kind of research in the drug development process qualified for the exemption:

basic research, preclinical research, or clinical studies. These three stages of drug development are described as follows: basic research involves the testing of thousands of compounds to discover any biological activity relevant to understanding the cause of a disease; the preclinical stage involves more focused research on a smaller group of chemical compounds in the hopes of finding the best candidate for clinical development; and clinical studies are the testing of the drug on human subjects in preparation for FDA approval.[21] Following its interpretive lead in *Eli Lilly,* the Supreme Court in *Merck KGaA v. Integra Lifesciences I, Ltd. ("Integra")* ruled that § 271(e)(1) immunizes from infringement both preclinical and clinical use of patented inventions in the drug research and development process.

MERCK KGAA V. INTEGRA LIFESCIENCES I, LTD.

Facts of the Case. Integra Lifesciences I, Ltd. ("Integra") is a pharmaceutical company that owns five patents related to a sequence of three amino acids, arginine, glycine, and aspartic acid (the "RGD peptide"), which promotes cell adhesion by attaching to receptors on cell surface proteins called integrins.[22] Scientists working for Telios Pharmaceuticals, Inc. discovered that the RGD peptide had potential use in promoting wound healing and biocompatibility of prosthetic devices, prompting Telios to obtain patents for the RGD peptide compositions and methods. However, after failing to develop a viable commercial product, Telios sold the patents to Integra.[23]

In the mid-1980s, Dr. David Cheresh at the Scripps Research Institute ("Scripps"), a non-profit corporation that conducts biochemical research, discovered that blocking integrin receptors using the RGD peptide inhibited angiogenesis, a process by which new blood vessels sprout from existing vessels. Angiogenesis plays a critical role in the spread of many diseases, including cancerous tumor growth, diabetic retinopathy, and rheumatoid arthritis.[24]

Merck KGaA ("Merck"),[25] a German pharmaceutical corporation unaffiliated with the U.S.-based pharmaceutical company Merck & Co., was interested in developing this discovery into a drug to control angiogenesis. In 1988, Merck entered into an agreement with Scripps to provide funding for Dr. Cheresh's research, in exchange for Scripps granting Merck an option to license future discoveries arising from his research.[26] In 1994, Dr. Cheresh succeeded in reversing tumor growth in chicken embryos using a RGD peptide identified as EMD 66203, which had been provided by Merck. This peptide was covered by Integra's patent.[27]

Due to Dr. Cheresh's breakthrough achievement, Merck and Scripps entered into a new collaboration agreement in September 1995 to fund the "necessary experiments to satisfy the biological bases and regulatory (FDA) requirements for the implementation of clinical trials" with EMD 66203 or a derivative thereof.[28] Dr. Cheresh then proceeded to conduct *in vivo* and *in vitro* experiments on EMD 66203 and two derivatives of it, EMD 85189 and EMD 121974, in order to evaluate each peptide as potential drug candidates. These "tests measured the efficacy, specificity, and toxicity of the particular peptides as angiogenesis inhibitors, and evaluated their mechanism of action and pharmacokinetics in animals."[29] Based on these tests, in November 1996 Merck's pharmaceutical steering committee selected EMD 85189 for pre-clinical development; in April 1997, Merck switched to EMD 121974 as its most promising candidate for clinical testing.[30] In October 1998, Merck reached an agreement with the National Cancer Institute (NCI) to sponsor the clinical trials, and later that year, the NCI filed an IND application with the FDA for EMD 121974.[31]

When Integra became aware of Merck's agreement with Scripps to conduct angiogenesis research for commercial purposes, Integra offered Merck the opportunity to purchase licenses to use its patented RGD peptides. In July 1996, after Merck had declined the offer, Integra sued Merck, Scripps, and Dr. Cheresh, seeking monetary damages for Merck's alleged patent infringement and a declaratory judgment against Scripps and Dr. Cheresh.[32] In defense, Merck asserted that its actions involving the RGD peptides came within the common-law research exemption[33] and the statutory safe harbor afforded by § 271(e)(1).

The District Court's Decision in *Integra*. At the conclusion of trial, the U.S. District Court for the Southern District of California dismissed Integra's claim for declaratory judgment and held that the common-law research exemption protected Merck's pre-1995 use of the RGD peptides.[34] However, the court found that a question of fact remained as to whether Merck's post-1995 actions fell within the scope of the § 271(e)(1) safe harbor. The district court instructed the jury that, for Merck to prevail on the safe harbor defense, it must prove by a preponderance of the evidence that it was objectively reasonable for the company to believe that "there was a decent prospect" that the experiments "would contribute, relatively directly," to the generation of information likely to be relevant to the drug approval regulatory process.[35]

The jury found Merck liable for infringing Integra's patents and that Merck had failed to show that § 271(e)(1) protected its post-1995 research activities. The jury awarded damages of $15 million in royalties. In response to post-trial motions, the district court dismissed Integra's suit against Scripps and Dr. Cheresh, but affirmed the jury's monetary award, explaining that there was substantial evidence to show that the connection between the experiments and FDA review was "insufficiently direct to qualify for the [§ 271(e)(1) exemption]."[36]

***Integra* in the Federal Circuit.** In June 2003, a divided panel of the Court of Appeals for the Federal Circuit affirmed the district court's determination as to liability but reversed the court's refusal to modify the damages award.[37] The panel majority found that safe harbor does not "reach any exploratory research that may rationally form only a predicate for future FDA clinical tests."[38] In confining the § 271(e)(1) exemption to research activities that contribute "relatively directly" to information "reasonably related" to clinical testing for the FDA, the appellate court stated:

> In this case, the Scripps work sponsored by Merck was not clinical testing to supply information to the FDA, but only general biomedical research to identify new pharmaceutical compounds. The FDA has no interest in the hunt for drugs that may or may not later undergo clinical testing for FDA approval.[39]

Furthermore, the court expressed concern that construing the safe harbor provision more expansively "would effectively vitiate the exclusive rights of patentees owning biotechnology tool patents," since patented research tools are often used in general research to identify candidate drugs and experiments on those drugs.[40]

On January 7, 2005, the U.S. Supreme Court granted *certiorari* to review the court of appeals' interpretation of the safe harbor provision.[41]

THE U.S. SUPREME COURT'S DECISION IN *INTEGRA*

The question presented to the Supreme Court was "whether uses of patented inventions in preclinical research, the results of which are not ultimately included in a submission to the Food and Drug Administration (FDA), are exempted from infringement by 35 U.S.C. § 271(e)(1)."[42] In a unanimous opinion written by Justice Scalia, the Court vacated the judgment of the Federal Circuit and held that the § 271(e)(1) safe harbor protected the preclinical use of patented compounds "as long as there is a reasonable basis for believing that the experiments will produce 'the types of information that are relevant to an IND or NDA'" submission to the FDA.[43]

The Court explained:

> [W]e think it apparent from the statutory text that § 271(e)(1)'s exemption from infringement extends to all uses of patented inventions that are reasonably related to the development and submission of *any* information under the FDCA. ... This necessarily includes preclinical studies of patented compounds that are appropriate for submission to the FDA in the regulatory process. There is simply no room in the statute for excluding certain information from the exemption on the basis of the phase of research in which it is developed or the particular submission in which it could be included.[44]

The Court rejected Integra's argument that the scope of the safe harbor is limited only to preclinical studies pertaining to the safety of a drug in humans.[45] Since the FDA requires an IND to be filed *before* human trials can begin, IND applications must include summaries of a drug's efficacy, pharmacokinetics, pharmacology, and toxicological effects in animals.[46] This data would necessarily have to be developed in preclinical studies — information that is "reasonably related" to an FDA submission and thus covered by § 271(e)(1).[47]

The Court further disagreed with Integra's claim that Merck's preclinical research is disqualified from safe harbor protection because the experiments were not conducted in conformity with the FDA's "good laboratory practices" (GLP) regulations. Two reasons supported the Court's reasoning: first, the FDA's GLP regulations concerning preclinical studies apply only to experiments on drugs "to determine their safety," and not to studies of a drug's efficacy, mechanism of action, pharmacology, or pharmacokinetics; second, even non-GLP compliant safety-related studies are suitable for submission in an IND, when such studies are accompanied by a reason for the noncompliance.[48]

Basic Research Not Protected. The Court placed an outer limit to the safe harbor provision by endorsing the Federal Circuit's conclusion that the exemption does not reach all experimental activity that at some point, however attenuated, may lead to an FDA approval process.[49] For example, safe harbor does not embrace basic scientific research performed on a patented compound without the intent to develop a particular drug or without a reasonable belief that the compound will cause a particular physiological effect that the researcher desires.[50] Thus, the boundary line between unprotected basic research and protected preclinical research is reached when a scientist discovers that a patented compound produces a "particular" physiological effect through a "particular" biological process.[51]

The Standard for "Reasonable Relation". In denying safe harbor protection for Merck's preclinical activities, the Federal Circuit had relied upon the fact that the "Scripps-Merck experiments did not supply information for submission to the [FDA], but instead identified the best drug candidate to subject to future clinical testing under the FDA processes."[52] The Supreme Court dismissed the appellate court's narrow interpretation of the "reasonably related" requirement in § 271(e)(1). Such a construction, the Court explained, "disregards the reality that ... scientific testing is a process of trial and error," and that "neither the drugmaker nor its scientists have any way of knowing whether an initially promising candidate will prove successful over a battery of experiments."[53] Thus, under certain conditions, the Court noted that the safe harbor provision is "sufficiently broad" to protect the use of patented compounds in experiments that are not ultimately submitted to the FDA or drug experiments that are not ultimately the subject of an FDA submission.[54]

The Court announced a standard for construing § 271(e)(1)'s reasonable relation requirement in a way that "leaves adequate space for experimentation and failure on the road to regulatory approval":

> At least where a drugmaker has a reasonable basis for believing that a patented compound may work, through a particular biological process, to produce a particular physiological effect, and uses the compound in research that, if successful, would be appropriate to include in a submission to the FDA, that use is "reasonably related" to the "development and submission of information under ... Federal law."[55]

An Unresolved Question: Patented Research Tools. Research tools are defined as "tools that scientists use in the laboratory, including cell lines, monoclonal antibodies, reagents, animal models, growth factors, combinatorial chemistry and DNA libraries, clones and cloning tools (such as PCR), methods, laboratory equipment and machines."[56] Smaller biotechnology companies and universities that invent research tools are concerned that a broader construction of § 271(e)(1) encompassing these tools will deprive them of licensing fees that they collect from larger pharmaceutical firms.[57] Moreover, some companies rely on such fees for their financial existence, since many of these research tools have little commercial value beyond usage in drug research.[58] The Federal Circuit in *Integra* had specifically identified this potential negative consequence for patented research tools, in its support for a more limited safe harbor:

> [T]he context of this safe harbor originally keyed its use to facilitating expedited approval of patented pioneer drugs already on the market. Extending § 271(e)(1) to embrace all aspects of new drug development activities would ignore its language and context with respect to the [Hatch-Waxman Act] in an attempt to exonerate infringing uses only potentially related to information for FDA approval. Moreover, such an extension would not confine the scope of § 271(e)(1) to *de minimis* encroachment on the rights of the patentee. For example, expansion of § 271(e)(1) to include the Scripps-Merck activities would effectively vitiate the exclusive rights of patentees owning biotechnology tool patents. Thus, exaggerating § 271(e)(1) out of context would swallow the whole benefit of the Patent Act for some categories of biotechnological inventions. Needless to say, the [Hatch-Waxman Act] was [not] meant ... to deprive entire categories of inventions of patent protection.[59]

In its amicus curiae brief submitted to the Supreme Court, the U.S. Government suggests that § 271(e)(1) does not apply to patented research tools.[60] The Government's brief explains that the safe harbor section, by its own terms, applies only to "a patented invention." The Patent Act defines the term "invention" to mean any "invention or discovery," "*unless* the context otherwise indicates."[61] The brief asserts that the context of § 271(e)(1) indicates that Congress may not have intended to include patented research tools within the scope of the safe harbor exemption. Since most research tools are used to study or develop other compounds for submission to the FDA regulatory approval process, rather than being themselves the subject of FDA regulatory review, it is plausible to conclude that research tools are not "patented inventions" within the meaning of the statute.[62]

In *Integra,* the Supreme Court expressly declined to decide whether or to what extent the exemption applies to patented research tools since the matter was not at issue in the case. The Court explained that Integra had never argued that the RGD peptides were used by Merck/Scripps as research tools, "and it is apparent from the record that they were not."[63] Thus, without a definitive judicial determination from the Court, the use of patented research tools in drug research and development may or may not fall under the § 271(e)(1) exemption from infringement. Such uncertainty over the patent rights of makers of research tools could serve as a source of continued confusion and litigation in this area.

CONCLUDING OBSERVATIONS

The original legislative intent behind the Hatch-Waxman Act that created § 271(e)(1) was to facilitate the introduction of a generic drug upon the patent expiration of the brand name drug. However, as the Supreme Court explained in the *Eli Lilly* case that broadened § 271(e)(1) beyond generic drugs to the entire statutory scheme of FDA regulation:"[I]t is not the law that a statute can have no effects which are not explicitly mentioned in its legislative history."[64]

The consequences of the Supreme Court's decision in *Integra* are significant. Some observers argue that if the Federal Circuit's opinion had not been vacated, its narrow interpretation of the patent law's safe harbor potentially would have created a chilling effect on the development of innovative, pioneer drugs and new generic drugs. Limiting § 271(e)(1) to only clinical research appears contrary to the objectives of the Hatch-Waxman Act: If a drug manufacturer could not perform the preclinical studies needed to obtain FDA approval to conduct clinical studies, "the [§ 271(e)(1)] exemption would never be reached because the underlying preliminary research and development work could not be undertaken" without risking patent infringement liability.[65]

The Supreme Court's more expansive construction of § 271(e)(1) avoids this result. Since "it will not always be clear to parties setting out to seek FDA approval for their new product exactly which kinds of information, and in what quantities, it will take to win that agency's approval," the safe harbor provision is needed to immunize certain preclinical studies that use patented compounds.[66] The Court also provided an articulated standard for courts, scientists, drug companies, and patent holders to follow concerning the scope of § 271(e)(1) coverage: Safe harbor applies if there is a reasonable basis to believe that the preclinical experiments will produce information that is relevant to an IND or NDA submission with the FDA. Failure to meet this standard would constitute infringing conduct not exempted by § 271(e)(1). By unanimous opinion, the *Integra* Court has emphatically clarified that preclinical use of patented compounds in pharmaceutical research is not categorically unprotected and can qualify for the patent law's safe harbor as long as it comes within this enunciated standard.

However, the *Integra* Court left unresolved the issue of whether research tools come within the scope of the safe harbor exemption. It is important to note that *Integra* does not affect the validity and value of patented research tools when they are employed in basic research or for purposes unrelated to an FDA submission.[67] Yet the unauthorized use of research tools in the development of information for the FDA regulatory process may constitute infringing conduct or could be exempted by the patent law's safe harbor. This legal uncertainty raises concerns about the enforceability of research tool patents in this circumstance. Unless or until the Supreme Court answers this question in a future case, Congress may desire to clarify § 271(e)(1)'s applicability to research tools.

REFERENCES

[1] Integra Lifesciences I, Ltd. v. Merck KGaA, 331 F.3d 860, 867 (Fed. Cir. 2003) (citation omitted).
[2] Merck KGaA v. Integra Lifesciences I, Ltd., __ U.S. __, 125 S. Ct. 2372, 2383 (2005).
[3] *Id.* at 2380.
[4] 35 U.S.C. § 271(a).
[5] 35 U.S.C. § 271(e)(1).
[6] P.L. 98-417, 98 Stat. 1585 (1984), codified in 15 U.S.C. §§ 8b-68c, 70b; 21 U.S.C. §§ 301, 355, 360cc; 28 U.S.C. § 2201; and 35 U.S.C. §§ 156, 271, 282.
[7] The statutory exemption is also called the "Bolar Amendment " or "FDA exemption," since it effectively overturns the decision of the U.S. Court of Appeals for the Federal Circuit in Roche Products, Inc. v. Bolar Pharmaceutical Co., Inc., 733 F.2d 858 (Fed. Cir. 1984), which had found Bolar, a manufacturer of generic drugs, liable for infringing Roche's patented drug during the last six months of the term of the patent in its testing and investigation activities related to FDA drug approval requirements.
[8] *Roche,* 733 F.2d at 860.
[9] For more information regarding the Hatch-Waxman Act, *see* CRS Report RL30756, *Patent Law and Its Application to the Pharmaceutical Industry: An Examination of the Drug Price Competition and Patent Term Restoration Act of 1984 ("The Hatch-Waxman Act"),* by Wendy H. Schacht and John R. Thomas; and CRS Report RL32377, *The Hatch-Waxman Act: Legislative Changes Affecting Pharmaceutical Patents*, by Wendy H. Schacht and John R. Thomas.
[10] P.L. 75-717 (1938), codified in 21 U.S.C. §§ 301 et seq.
[11] 21 U.S.C. § 355(i). Generic drug companies may file an abbreviated new drug application (ANDA) with the FDA. 21 U.S.C. § 355(j). An ANDA must reveal that the generic product has the same active ingredients as, and is bioequivalent to, a prior approved brand name drug. Also, in its ANDA, the generic drug manufacturer may rely upon the safety and efficacy data of the original drug manufacturer.
[12] 21 U.S.C. § 355(i)(1)(A).
[13] 21 U.S.C. § 355(b)(1).
[14] For more information concerning the FDA drug approval process, *see* CRS Report RL32797, *Drug Safety and Effectiveness: Issues and Action Options After FDA Approval,* by Susan Thaul, and CRS Report RL30989, *The U.S. Drug Approval Process: A Primer*, by Blanchard Randall IV.
[15] H.Rept. 98-857 (II) at 30, 98th Cong., 2d Sess. (1984), *reprinted in* 1984 U.S.C.C.A.N. 2686, 2714.

[16] 496 U.S. 661 (1990).
[17] *Id.* at 665.
[18] *Id.* at 667.
[19] *Id.* at 666.
[20] *Id.* at 674.
[21] James N. Czaban & Nishita Doshi, *Supreme Court Applies Broad Interpretation of Bolar Amendment to Protect Innovative Drug Research From Claims of Patent Infringement,* 70 PAT., TRADEMARK, & COPYRIGHT J. (BNA) 1726 (June 24, 2005).
[22] *Integra*, 331 F.3d at 862-63.
[23] *Id.* at 873 (Newman, J., dissenting).
[24] *Id.* at 863.
[25] *See* Merck is Not the Same as Merck, *available at* [http://www.merck.de/servlet/PB/menu/1014710/index.html].
[26] Telios Pharms., Inc. v. Merck KGaA, 1997 U.S. Dist. LEXIS 24187, Case No. 96-CV-1307 (S.D. Cal., Sept. 9, 1997), at *3.
[27] *Integra,* 125 S. Ct. at 2378 n.3.
[28] *Integra,* 331 F.3d at 863.
[29] *Integra,* 125 S. Ct. at 2378. Efficacy means how well a drug can be expected to work in curing a disease; mechanism of action is how it achieves those results; pharmacokinetics is the rate at which a drug is absorbed into and eliminated from the bloodstream; and toxicity is the negative side effects of the drug at different dosages. Brief for Petitioner Merck KGaA at 12-13, *Merck KGaA v. Integra Lifesciences I, Ltd.,* 125 S. Ct. 2372 (2005) (No. 03-1237).
[30] *Telios Pharms.,* 1997 U.S. Dist. LEXIS 24187, at *5.
[31] *Integra,* 331 F.3d at 873 (Newman, J., dissenting).
[32] *Id.* at 863.
[33] The common-law research exemption is a limited, judge-made exception to the patentee's right to exclude. Its historic foundations arise from Whittemore v. Cutter, 29 Fed. Cas. 1120, 1121 (C.C.D. Mass. 1813), in which Justice Story stated, "it could never have been the intention of the legislature to punish a man, who constructed such a machine merely for philosophical experiments..." Courts have recognized the use of this exemption for research that has no commercial purpose. *Integra,* 331 F.3d at 874-75 (Newman, J., dissenting).
[34] *Integra,* 125 S. Ct. at 2379.
[35] *Id.*
[36] *Id.* at 2380.
[37] On remand, the District Court reduced the award to $6.375 million, on the calculated basis of $1.5 million per year as a reasonable royalty between the infringement period August 1994 and November 1998. *Integra*, 2004 WL 2284001, at *11 (S.D. Cal. Sept. 7, 2004).
[38] *Integra,* 331 F.3d at 867.
[39] *Id.* at 866-67.
[40] *Id.* at 868.
[41] *Integra,* 125 S. Ct. 823 (2005).
[42] *Integra,* 125 S. Ct. at 2376.
[43] *Id.* at 2383-84 (citation omitted).

[44] *Id.* at 2380 (emphasis in original) (citations omitted).
[45] *Id.* at 2381 ("We do not understand the FDA's interest in information gathered in preclinical studies to be so constrained.")
[46] 21 C.F.R. § 312.23(a)(5).
[47] *Integra,* 125 S. Ct. at 2381.
[48] *Id.* at 2382 (citations omitted).
[49] *Id.*
[50] *Id.*
[51] *Id.* at 2383.
[52] *Integra,* 331 F.3d at 865-66.
[53] *Integra,* 125 S. Ct. at 2383.
[54] *Id.* at 2382. The legislative history of § 271(e)(1) supports this reasoning: "A party which develops [information for the FDA regulatory process], but decides not to submit an application for approval, is protected [by the safe harbor] as long as the development was done to determine whether or not an application for approval would be sought." H.Rept. 98-857 (I) at 45, 98th Cong., 2d Sess. (1984), *reprinted in* 1984 U.S.C.C.A.N. 2647, 2678.
[55] *Integra,* 125 S. Ct. at 2383 (citing § 271(e)(1)).
[56] *Integra,* 331 F.3d at 874 n.4 (citing *Sharing Biomedical Research Resources: Principles and Guidelines for Recipients of NIH Research Grants and Contracts,* 64 Fed. Reg. 72090, 72092 n.1 (Dec. 23, 1999)).
[57] David Savage & Denise Gellene, *High Court Boosts Drug Research; Justices Say Companies Are Free to Use Patented Compounds in Developing Medicines. Analysts Say Ruling May Hurt Some Biotech Firms,* L.A. TIMES, June 14, 2005, at C1.
[58] *Id.*
[59] *Integra,* 331 F.3d at 867.
[60] Brief for United States as *Amicus Curiae*, at 29, *Merck KGaA v. Integra Lifesciences I, Ltd.,* 125 S. Ct. 2372 (2005) (No. 03-1237).
[61] 35 U.S.C. § 100(a) (emphasis added).
[62] Brief for United States as *Amicus Curiae*, at 29.
[63] *Integra,* 125 S. Ct. at 2382 n.7.
[64] *Eli Lilly,* 496 U.S. at 669 n.2 (quoting Pittston Coal Group v. Sebben, 488 U.S. 105, 115 (1988)).
[65] Brief for United States as *Amicus Curiae*, at 14 (quoting Bristol-Myers Squibb Co. v. Rhone-Poulenc Rorer, Inc., 2001 WL 1512597, at *6 (S.D.N.Y. Nov. 28, 2001)).
[66] *Integra,* 125 S. Ct. at 2383 (citation omitted).
[67] Brief for United States as *Amicus Curiae*, at 30.

In: New Intellectual Property Issues
Editor: John V. Luong, pp. 129-143

ISBN: 978-1-60021-211-6
© 2007 Nova Science Publishers, Inc.

Chapter 7

ANTICIRCUMVENTION UNDER THE DIGITAL MILLENNIUM COPYRIGHT ACT AND REVERSE ENGINEERING: RECENT LEGAL DEVELOPMENTS*

Robin Jeweler

SUMMARY

The Digital Millennium Copyright Act (DMCA) prohibits individuals from manufacturing, selling, or trafficking in technology, products, services, or devices that circumvent technology designed to control access to a copyrighted work. This is known as the DMCA's "anticircumvention" provision. Although most commonly invoked in the context of digital piracy of music, motion pictures, and other entertainment-related media, another genre of anticircumvention-based cases is making its way through the courts. These cases involve the initiation of anticircumvention litigation for what some argue are anti-competitive purposes in the marketing and sale of durable goods.

The practice of reverse engineering allows others to identify and analyze the creative versus the functional aspects of copyrighted software and to utilize them to some degree. In some contexts, reverse engineering has been held by the courts to be a fair use of copyright-protected property. There is also an express, limited statutory exemption for reverse engineering under the DMCA. How the practice and use of reverse engineering for commercial goals relates to the relatively new protections against circumvention is of interest to many. They are concerned with the extent to which access-control technology is likely be employed to extend (or attempt to extend) a copyright holder's control over durable goods with copyrighted components and the secondary markets for such goods.

This report examines two recent decisions from U.S. Courts of Appeals, *Lexmark International, Inc. v. Static Control Components, Inc.*, 387 F.3d 522 (6th Cir. 2004), and *Chamberlain Group, Inc. v. Skylink Technologies, Inc.*, 381 F.3d 1178 (Fed. Cir. 2004). Both cases allege violation of the anticircumvention statute with respect to development and sale of consumer goods. In *Lexmark*, the defendant marketed a microchip that allowed third-party manufacturers to sell toner cartridges that worked with the plaintiff's printer. In *Chamberlain*, the defendant sold a universal garage door opener transmitter that worked with the plaintiff's garage door opener. In both cases, the courts found that the actions of the defendants did *not*

---

* Excerpted from CRS Report RL32692, Updated December 10, 2004

violate the anticircumvention provisions of the DMCA. In doing so, the courts had to reconcile closely related aspects of copyright law with the strictures of the DMCA.

BACKGROUND

The Digital Millennium Copyright Act (DMCA) prohibits individuals from manufacturing, selling, or trafficking in technology, products, services, or devices that circumvent technology designed to control access to a copyrighted work.[1] This is known as the DMCA's "anticircumvention" provision. Enacted in 1998, the law became effective in 2000. The DMCA's anticircumvention provisions have proven to be controversial. Proponents argue that they are essential to protect the creation and distribution of, and the market for digital intellectual property (IP), while critics argue that they extend the copyright monopoly beyond its intended scope and thwart the public's ability to access materials for permissible purposes.[2]

The statute is best known in the context of encryption programs and access control "gates" interposed between the public and copyrighted digital entertainment media, e.g., music and motion pictures.[3] Here, critics of the DMCA argue that the anticircumvention provisions allow copyright holders to impose new and overly restrictive conditions on content users. This new layer of IP protection has been referred to as "paracopyright."[4] It refers to the ability of content owners to use conditions for access in order to extend control over usage of copyrighted work.

The focus of this report, however, is another genre of anticircumvention-based cases making their way through the courts. These cases involve the initiation of anticircumvention litigation for what some argue are anti-competitive purposes in the marketing and sale of durable goods. The question is, when a manufacturer utilizes software as a "gate" to enable, control, or restrict access to consumer goods containing copyrighted material, to what extent does it preclude competitors from circumventing the controls in the engineering of compatible – or competitive –consumer goods?

This report examines two cases involving 17 U.S.C. § 1201 that involve commercial interests unrelated to the entertainment industry, specifically, manufacturers of garage door openers and manufacturers of toner cartridges for photocopying machines. These cases involve issues that have been the basis for criticism of the DMCA in a slightly different commercial context, namely, whether the provisions permit IP owners to extend their copyright monopoly beyond its intended scope to control consumer goods.

EXCEPTIONS TO THE COPYRIGHT MONOPOLY: REVERSE ENGINEERING

There are several statutory and judicially created exceptions to a copyright holder's right to control IP. Among the best known is the public's right to make "fair use" of a protected work.[5] Among the judicially-sanctioned interpretations of the fair use defense to copyright infringement is the right to "reverse engineer" a copyrighted computer program to gain an understanding of its unprotected functional elements.[6] Like all fair use analyses, the context for a court's decision is intensely fact-specific. But a court considering the legitimacy of reverse engineering as a fair use will factor in the public interest in a competitive marketplace for consumer goods. When, for example, copyrighted computer code functions primarily as a "lock out" code designed to prevent access to functional as opposed to copyrighted elements of a durable good, then reverse engineering or other methods of overcoming it may be permissible.[7]

In *Sega Enterprises Ltd. v. Accolade, Inc.*, [8] a U.S. Court of Appeals held that Accolade, Inc.'s disassembly and reverse engineering of Sega's copyrighted computer program was a fair use. Accolade reverse engineered the program in order to manufacture video games on cartridges that were compatible with Sega's video console, Genesis. Because there was no other known method of access, the court found that Accolade's use of Sega's initialization code did not violate the U.S. Copyright Act. Accolade's copies of Sega's software enabled it to discover the functional requirements for compatibility with the Genesis console. Even though Accolade's ultimate purpose was the development of Genesis-compatible games for sale, its direct use of the copyrighted material and its purpose in reverse engineering Sega's code was to study the *functional* requirements for Genesis compatibility.

Accolade wanted to modify its existing games so that they could be played on the Genesis console. In reaching its decision, the court noted that

> [W]e are free to consider the public benefit resulting from a particular use notwithstanding the fact that the alleged infringer may gain commercially. Public benefit need not be direct or tangible, but may arise because the challenged use serves a public interest. In the case before us, Accolade's identification of the functional requirements for Genesis compatibility has led to an increase in the number of independently designed video game programs offered for use with the Genesis console. It is precisely this growth in creative expression, based on the dissemination of other creative works and the unprotected ideas contained in those works, that the Copyright Act was intended to promote.[9]

Judicial determination of what activity constitutes a fair use of copyrighted property is a fluid and evolving doctrine. The anticircumvention provisions of the DMCA, however, establish offenses wholly distinct from copyright infringement. In other words, reverse engineering that may be (or have been) permissible under the Copyright Act may no longer be permissible under the DMCA. To the extent that the DMCA's anticircumvention provisions address the narrow issue of manipulating – or circumventing – technology (whether or not it is copyrightable in its own right) *to access* copyrighted technology, its impact on fair use, particularly the employment of reverse engineering, may have a significant impact in the marketplace.

And, there are still many other legal doctrines that impact the legality of reverse engineering in any given context. For example, the DMCA itself, at 17 U.S.C. § 1201(f), contains an express exemption for reverse engineering "for the sole purpose of identifying and analyzing those elements of the program that are necessary to achieve interoperability of an independently created computer program with other programs[.]"[10] Conversely, the courts have honored contractual terms under state law, such as licensing agreements and terms of use agreements, that curtail a user's right to reverse engineer copyrighted material.[11] How do the rights, remedies, and exceptions in the DMCA relate to comparable ones in the Copyright Act?

In the realm of online entertainment, such as music, film, and electronic books, critics of the DMCA argue that it has the potential to curtail the public's right to make fair use of copyrighted material. In other commercial contexts, litigants argue that the act's anticircumvention provisions may thwart access to the permissible fair use of reverse engineering and thereby permit a copyright holder to leverage a limited monopoly to control areas outside of that monopoly in the broader world of fungible consumer goods, a practice traditionally held in check by the judicial doctrine of "copyright misuse."[12] But with the "rebalancing" of interests that was effected by enactment of the DMCA, how far can a manufacturer go in using restrictions on access control technology to prevent competitors from legitimately using reverse engineering to develop competitive, interoperable noninfringing consumer goods?[13] This question, and the relationship between prohibitions on circumvention and the underlying law of copyright, makes the anticircumvention cases presently wending their way through the federal courts of interest to many.

LEXMARK INTERNATIONAL V. STATIC CONTROL COMPONENTS, INC.

A U.S. district court granted Lexmark, a manufacturer of laser printers and printer toner cartridges, a preliminary injunction against a remanufacturer of replacement cartridges, Static Control Components (SCC), for copyright infringement and violation of § 1201 of the DMCA.

A toner cartridge is a device that is inserted within a laser printer and contains the toner necessary for the printer to print. Plaintiff Lexmark manufacturers two types of toner cartridges: a regular one that can be refilled by the printer owner or by a third party remanufacturer and a discounted one for its T-series printers which, pursuant to a licensing agreement, consumers agree to use only once and return to Lexmark for recycling. SCC and other remanufacturers refill and sell Lexmark's regular toner cartridges.

Lexmark employs copyrighted computer programs embedded within microchips to control and monitor various operations on the T-series toner cartridges. It utilizes an authentication sequence to prevent unauthorized access to its Printer Engine and Toner Loading Program (TLP). In the absence of the authentication sequence, a printer will not recognize a toner cartridge as being authorized and disables access to the Printer Engine Program.

Defendant SCC manufactures and sells components for use in the remanufacturing of toner cartridges. By reverse engineering the Lexmark microchip, SCC developed a microchip that copies, and thereby circumvents, the authentication sequence that Lexmark uses in the T-Series printers. This enables SCC to sell its "SMARTEK" microchip for use by third-parties to market replacement toner cartridges that are compatible with the T-series despite Lexmark's intention to limit replacement cartridges to those it manufactures.

U.S. District Court

Lexmark sued SCC in U.S. district court and received a preliminary injunction against the continuing sale by SCC of the "SMARTEK" microchip for T-series replacement cartridges. The court found that Lexmark was likely to prevail on claims of copyright infringement and violation of the DMCA.[14]

There were two distinct components of SCC's microchip. One circumvented an authentication sequence embedded in the T-series printer. The sequence determines whether the printer will accept the cartridge as authorized and subsequently grant it access to the Printer Engine Program which is resident in the printer's controller board. The other both circumvented access to and copied Lexmark's TLP in its entirety. The Toner Loading Program enables the printers to approximate the amount of toner remaining in the cartridge. This information is used to display a "toner low" message on the printer screen.

The question before the court was twofold: First, did the SMARTEK microchip violate the DMCA by illegally circumventing the access control systems to both the Printer Engine Program and the Toner Loading Program? Second, by replicating the TLP in its entirety, did the microchip infringe Lexmark's copyright in it? The court considered in depth whether the Lexmark's Toner Program was copyrightable and concluded that it was, relying in part upon the fact that the U.S. Copyright Office granted Lexmark of a Certificate of Registration for it. It considered and rejected a fair use defense. It also rejected a copyright misuse defense, finding that attempting to enforce rights under the DMCA to protect access to a copyrighted computer program *cannot* have the legal effect of "using copyright to secure an exclusive right or limited monopoly not expressly granted by copyright law."[15]

Nor could Lexmark's efforts to enforce its rights under the DMCA be considered an unlawful act undertaken to stifle competition. The court emphasized that under the DMCA, "the right to protect against unauthorized access is a right separate and distinct from the right to protect against violations of exclusive copyright rights such as reproduction and distribution."[16] Indeed, the authentication sequence which triggered the interface between Lexmark's toner cartridge microchip and printer was exactly the type of "technological measure that effectively controls access to a work" designed to be protected under the DMCA.[17]

Finally, the court rejected the argument that the DMCA's exemption for reverse engineering was applicable to SCC's SMARTEK microchip.[18] In addition to violating the DMCA by circumventing the authentication sequence to the Printer Engine Program, SCC's microchip infringed the Copyright Act because it contained an exact copy of Lexmark's copyrighted Toner Loading Program. Therefore, SCC was unable to meet the reverse engineering exemption requirement that the circumvention enables interoperability of an *independently created* computer program.

The Court of Appeals

The Sixth Circuit Court of Appeals overturned the lower court's issuance of a preliminary injunction, taking issue with virtually all of the district court's findings.[19] Specifically, the Court of Appeals found that, for purposes of upholding a preliminary injunction, Lexmark was not likely to successfully establish that SCC infringed its copyright for the TLP or that the SMARTEK microchip violated the DMCA by illegally circumventing access controls.

The court first considered the TLP's eligibility for copyright protection. It weighed the basic element for copyrightability, original and/or creative expression, against that which *cannot* be copyrighted – an idea, procedure, process, system, method of operation, concept, principle or discovery, regardless of form.[20] Applying this process to computer programs with its concomitant task of separating idea from expression is "vexing."[21] In order to ascertain the "elusive boundary line" between idea/expression and between process/non-functional expression, the court utilized the doctrines of "merger" and "scènes à faire." The merger doctrine precludes copyright protection where there is only one way or very few ways of expressing an idea. Idea and expression are deemed to be merged because granting copyright to the expressive component of the work would extend protection to the work's uncopyrightable ideas as well. The scènes à faire doctrine is related. In the computer-software context it means

> that the elements of a program dictated by practical realities--*e.g.*, by hardware standards and mechanical specifications, software standards and compatibility requirements, computer manufacturer design standards, target industry practices, and standard computer programming practices--may not obtain [copyright] protection.[22]

The court noted that lock-out codes generally fall on the functional-idea rather than the original-expression side of the copyright line. Manufacturers of interoperable devices such as toner cartridges or garage door opener transmitters often employ a security system to bar the use of unauthorized components. To unlock the system to permit operation of the primary device, the printer or the garage door opener, the component must contain certain code or be engineered to respond correctly to an authentication sequence. When a code sequence must be included in a component device to permit its use, the merger and scènes à faire doctrines often preclude the sequence from obtaining copyright protection.

Applying the foregoing, the court concluded that the district court erred in finding that the TLP code had sufficient originality to be copyrightable. It found the TLP to be a very brief, uncomplicated code sequence that functioned as a lock out code. Brevity in a computer program does not make it ineligible for copyright protection, but creativity and effort are reciprocally related: the smaller the effort, the greater the creativity required to invoke copyright protection.[23]

In view of its finding that the TLP was not copyrightable, the court declined to rule on SCC's fair use defense. It did, however, indicate that although the purpose of SCC's use of the SMARTEK chip was for commercial gain, a factor in a fair use analysis, its use of the copyrighted TLP on the chip was not. Rather, the use of the TLP was to satisfy authentication requirements and to permit printer functionality. Likewise, with respect to the allegedly infringing use on the value of the copyrighted work, the lower court had concluded that the SMARTEK chip would diminish the market for sales of Lexmark's toner cartridges. The proper and more narrow question, however, was whether replication of the TLP on the microchip would diminish demand for the TLP. More significantly, the Court of Appeals observed that copyright law is not properly invoked to protect a secondary market in consumer goods:

> [T]he district court focused on the wrong market: it focused not on the value or marketability of the Toner Loading Program, but on Lexmark's market for its toner cartridges. Lexmark's market for its toner cartridges and the profitability of its Prebate program may well be diminished by the SMARTEK chip, but that is not the sort of market or value that copyright law protects. ... Lexmark has not introduced any evidence showing that an independent market

exists for a program as elementary as its Toner Loading Program, and we doubt at any rate that the SMARTEK chip could have displaced any value in *this* market.[24]

The Court of Appeals then addressed the district court's findings with respect to Lexmark's likelihood of success in establishing that SCC violated the DMCA's ban on distributing devices that circumvent access-control measures placed on copyrighted works. According to Lexmark, SCC's SMARTEK chip is a "device" that "circumvents" Lexmark's "technological measure," i.e., the authentication sequence, which "effectively controls access" to its copyrighted works (the Toner Loading Program and Printer Engine Program). Lexmark claimed that the SMARTEK chip met all three tests for liability under § 1201(a)(2): (1) the chip was "primarily designed or produced for the purpose of circumventing" Lexmark's authentication sequence, (2) it had "only limited commercially significant purpose or use other than to circumvent" the authentication sequence; and (3) SCC marketed the chip "for use in circumventing" the authentication sequence. The district court agreed and concluded that Lexmark had shown a likelihood of success under all three provisions.

For the Court of Appeals, the anticircumvention allegation with respect to the TLP was problematic. The microchip did not provide "access" to the TLP but actually replaced it. More important, since the DMCA prohibits circumvention of access controls for *copyrighted* work, the determination that the TLP was not eligible for copyright protection mooted the DMCA charge.

The charge of illegal circumvention of the authentication sequence for the copyrighted Printer Engine Program was more complicated. The district court determined that the authentication sequence "controlled access" to the Printer Engine Program because it controlled the consumer's "ability to make use of" the program. The Court of Appeals disagreed, asserting that the authentication sequence did *not* control access, rather "[i]t is the purchase of a Lexmark printer that allows <access' to the program."[25] The literal code itself can be read directly from the printer memory without the benefit of the authentication sequence. There is no security device protecting the Printer Engine Program. While the authentication sequence controls access to the Printer Engine Program's functionality, the Program itself is not otherwise effectively protected by access controls; the data and program can be translated into readable source code. Hence, the authentication sequence controlled only one avenue of access – operational compatibility with a printer cartridge. Analogizing access controls to locks on a house, the court reasoned:

Because the statute refers to "control[ling] access to a work protected under this title," it does not naturally apply when the "work protected under this title" is otherwise accessible. Just as one would not say that a lock on the back door of a house "controls access" to a house whose front door does not contain a lock and just as one would not say that a lock on any door of a house "controls access" to the house after its purchaser receives the key to the lock, it does not make sense to say that this provision of the DMCA applies to otherwise-readily-accessible copyrighted works.[26]

The court elaborated by explaining why access to the Printer Engine Program is not covered by the DMCA. In the setting where the DMCA applies, such as encrypted data on music CDs and motion-picture DVDs, the copyright content that it protects operates on "two planes": the literal code governing the content and its manifestation generated by the code's execution, namely, the music or movie. In the DMCA's context, restricting "use" of the work means "restricting consumers from making use of the copyrightable expression in the work."[27] The manufacturer must prevent access to both planes of the copyrighted material and the alleged infringer responds by marketing a device that circumvents both levels of protection. Thus, because Lexmark did not direct any of its security efforts, through the authentication sequence or otherwise, to ensuring that the Printer Engine Program could not be read and copied, "it cannot lay claim to having put in place a <technological measure that effectively controls access to a work protected under [the copyright statute].'"

The extent to which the court leaves open an invitation to manufacturers to come within the DMCA by withholding all levels of access to copyright-protected code is less clear. It refers to the act's legislative history and discerns congressional intent to protect the market for interoperable consumer goods, but nevertheless emphasizes its "two plane" analysis:

> Nowhere in its deliberations over the DMCA did Congress express an interest in creating liability for the circumvention of technological measures designed to prevent consumers from using consumer goods while leaving the copyrightable content of a work unprotected. In fact, Congress added the interoperability provision in part to ensure that the DMCA would not diminish the benefit to consumers of interoperable devices "in the consumer electronics environment." 144 Cong. Rec. E2136 (daily ed. Oct. 13, 1998) (remarks of Rep. Bliley).[28]

Interestingly, in a concurring opinion, Circuit Judge Merritt addresses this very point. He advocates broadening the scope of the holding by making clear "that in the future companies like Lexmark cannot use the DMCA in conjunction with copyright law to create monopolies of manufactured goods for themselves just by tweaking the facts of this case: by, for example, creating a Toner Loading Program that is more complex and <creative' than the one here, or by cutting off other access to the Printer Engine Program."[29]

In his opinion, the critical question is the purpose of the circumvention technology. Because the purpose of the law is to prohibit piracy of copyright protected works such as movies, music, and computer programs, an anticircumvention claim should not be permitted to go forward unless the plaintiff can demonstrate that the circumvention supports piracy. A broad reading of the statute is necessary, he argues, to prevent manufacturers from creating monopolies for durable goods such as replacement parts through the use of more elaborate lock out codes. And the burden of proof to demonstrate piracy must be on the plaintiff in order to prevent the threat of litigation by powerful manufacturers against smaller rivals and thereby thwart development of devices that facilitate legitimate access to consumer goods. Only then should the burden of proof shift to the defendant to invoke a statutory exception, such as the reverse engineering exception.

CHAMBERLAIN GROUP, INC. V. SKYLINK TECHNOLOGIES, INC

In three reported decisions, a U.S. district court and the Court of Appeals for the Federal Circuit considered the claim of plaintiff Chamberlain, a garage door opener (GDO) manufacturer, that defendant Skylink's universal remote control GDO transmitter violated the DMCA. Chamberlain manufactures and sells a variety of GDO systems. Its "Security+" line incorporates a copyrighted computer program known as its "rolling code" system to transmit changing signals. The rolling code system is designed to enhance the security of its GDO by preventing "code grabbers" from intercepting the GDO signal. Skylink markets a universal remote transmitter, Model 39, which functions with a wide variety of GDOs. It does not incorporate Chamberlain's rolling code, but has one specific setting that operates the rolling code GDO. Chamberlain sued Skylink, asserting that the Model 39 transmitter is marketed for use in circumventing its rolling code computer program and that it renders the Security+ GDO insecure by allowing unauthorized users to circumvent the security inherent in the rolling code. Consequently, Chamberlain contended that Skylink violated the anti-trafficking clause of the DMCA's anticircumvention provision, § 1201(a)(2).

U.S. District Court Holdings

In the first of two reported U.S. district court decisions, the court considered whether Chamberlain's rolling code security system was protected by copyright and whether the DMCA applied.[30] The court in denying Chamberlain's motion for summary judgment established a framework for its anticircumvention analysis, noting that it is "undisputed" that the DMCA's application is not limited to the Internet.[31] It was specifically intended to prohibit the trafficking of products or devices that circumvent technological measures to restrict access to copyrighted work.

In the second opinion,[32] the court granted defendant Skylink's motion for summary judgment determining that the Model 39 transmitter did not provide *unauthorized* access to Chamberlain's copyrighted software. Skylink's program to overcome Chamberlain's rolling code did not satisfy the statutory definition of circumvention *without the authority of the copyright owner*.[33] The court did not require Skylink to prove that its use was authorized. Rather, the burden was on Chamberlain to demonstrate that the use was unauthorized. The court found the following criteria persuasive to support its conclusion that Skylink's use of the rolling code was authorized:

- Chamberlain did not put any restrictions on consumers regarding the type of transmitter they needed buy to operate its GDO;
- Chamberlain also marketed a universal transmitter and did not advise consumers that they were limited to purchasing replacements from Chamberlain; and
- In order for a Skylink transmitter to operate a Chamberlain GDO, the homeowner must store the transmitter's signal into the GDO's memory. Hence, it is the consumer, not Skylink, that authorizes access.

Skylink's Model 39 universal GDO transmitter did not violate the statute because creating and marketing one that is interoperable with the Chamberlain GDO was implicitly authorized.

The court emphasized two factors: homeowner expectations and industry practice. In response to Chamberlain's argument that it never warned customers against using unauthorized transmitters because it had no idea that other transmitters could be made to operate its rolling code, the court replied that Chamberlain's failure to anticipate new technology "does not refute the fact that homeowners have a reasonable expectation of using the technology now that it is available."[34] The fact that there is "a history in the GDO industry of universal transmitters being marketed and sold to allow homeowners an alternative means to access any brand of GDO" negated Chamberlain's contention that there was an implied restriction against the use of competing transmitters.[35]

The court compared the absence of an explicit limitation by Chamberlain on the use of competing transmitters to Lexmark's explicit restrictions against use of third-party toner cartridge refills in its shrinkwrap agreement with consumers. Yet, while simultaneously acknowledging the limitations that may be effected by a shrinkwrap agreement, the court also suggested that "legitimate consumer expectations" may constitute an independent "fair use" basis for circumventing access control technology:

> [A] homeowner has a legitimate expectation that he or she will be able to access the garage even if the original transmitter is misplaced or malfunctions. .... Under Chamberlain's theory, any customer who loses his or her Chamberlain transmitter, but manages to operate the opener either with a non-Chamberlain transmitter or by some other means of circumventing the rolling code, has violated the DMCA. In this court's view, the statute does not require such a conclusion. GDO transmitters are similar to television remote controls in that consumers of both products may need to replace them at some point due to damage or loss, and may program them to work with other devices manufactured by different companies. In both cases, consumers have a reasonable expectation that they can replace the original product with a competing, universal product without violating federal law.[36]

The court assigned significant weight to two factors in issuing the summary judgment in favor of Skylink. First, that the process by which the Model 39 transmitter passed Chamberlain's rolling code was not created without authorization and, second, that the burden of proof to demonstrate the absence of authorization is on the party alleging violation of the anticircumvention statute. It therefore rejected Chamberlain's contention that Skylink violated the prima facie requirement of anti-trafficking under § 1201(a)(2) because its transmitter bypasses Chamberlain's rolling code security measure to gain access to Chamberlain's copyrighted GDO receiver operating software.

Court of Appeals

The Court of Appeals noted that to determine whether Skylink was entitled to summary judgment on the grounds that its Model 39 universal transmitter does not violate the DMCA requires the statutory construction of what § 1201(a)(2) does prohibit, a matter of "first impression."[37] It reviewed proceedings before the lower court, emphasizing the significance of the fact that Chamberlain did *not* allege that Skylink had infringed its copyright or was liable for contributory copyright infringement. And, with respect to the question of *unauthorized* circumvention, the court reiterated that the homeowner who purchases a Chamberlain GDO owns it and has a right to access his or her garage with it. Chamberlain did not place any explicit terms or conditions to limit the ways a purchaser may use its products. The homeowner who wishes to use a Skylink transmitter to operate with the Chamberlain GDO must program it to do so. Hence, if Chamberlain's interpretation of the DMCA were correct, not only would Skylink be in violation of the DMCA for trafficking in circumvention devices under § 1201(a)(2), but owners of a Chamberlain GDO who purchased a Skylink transmitter would be in violation of (a)(1) which prohibits circumvention.[38]

But the Court of Appeals, departing from the lower court's analysis, undertook a more expansive interpretation of the anticircumvention statute and did not limit its examination to the question of "authorization" as a component of the statutory definition of circumvention. It began by examining the "essence"of § 1201 which, it found, does not create a new property right. Circumvention is not legally synonymous with infringement:

> This distinction between property and liability is critical. Whereas copyrights, like patents, are property, liability protection from unauthorized circumvention merely creates a new cause of action under which a defendant may be liable. The distinction between property and liability goes straight to the issue of authorization, the issue upon which the District Court both denied Chamberlain's and granted Skylink's motion for summary judgment.[39]

The distinction is evident in the basis for stating the claim. A plaintiff alleging copyright infringement need only prove ownership and copying to make a claim. The burden of proving that the use was authorized falls on the defendant. But a plaintiff alleging illegal circumvention or trafficking must prove that the defendant's access was unauthorized, a significant burden of proof which requires the plaintiff to establish that copyright law does *not* permit the access. Only then does the burden shift to the defendant.

The court rejected Chamberlain's argument that enactment of the DMCA renders pre-DMCA history in the GDO industry "irrelevant" and overrides preexisting consumer expectations about the legitimate uses of products containing copyrighted-embedded software. It rejected the broader contention that, in the absence of explicit authorization, circumvention of any technological access control measure to use a product containing copyrighted software is *per se* illegal. To subscribe to such an interpretation of the statute would grant manufacturers "broad exemptions from both the antitrust laws and the doctrine of copyright misuse."[40]

In other words, the DMCA's prohibition against circumvention does create a new and independent cause of action separate from copyright infringement.[41] But in order to establish liability for unauthorized circumvention, the complainant must demonstrate that access is indeed linked to copyright infringement. And this linkage allows the court to consider all non-infringing uses, including statutorily and judicially established fair uses. Hence, to violate the DMCA by circumventing access technology "without the authority of the copyright owner" means to circumvent *for a purpose impermissible under the copyright laws*; not simply to circumvent without the express permission of the copyright owner.[42]

The court examined § 1201's construction and legislative history and determined that the key to disentangling these relationships lies in understanding the linkage between "access" and "protection." Tying prohibitions against access to protectible rights under copyright law represents a "rebalancing of interests" that addresses digital piracy. Granting copyright owners unqualified control over technological access would create "two distinct copyright regimes." The first regime permits copyright owners to protect only the rights enumerated in 17 U.S.C. § 106, subject to the exceptions and limitations of the Copyright Act. Owners who attempt to protect their rights by incorporating technological measures to protect against encroachment are able to hold traffickers in circumvention devices liable under § 1201(b) for putting their rights at risk by enabling users of the devices *to infringe*.

Under the second regime – inherent in Chamberlain's argument and rejected by the court – the owners of a work protected by both copyright *and* a technological measure that controls access to that work would possess *unlimited* rights to hold circumventors liable under 1201(a) simply for accessing that work, even if that access enabled *only* rights that the Copyright Act grants to the public. This interpretation, even under the "substantial deference due Congress ... borders on the irrational."[43]

Addressing concerns of many DMCA critics, the court explained that the doctrines of antitrust and copyright misuse are inescapable components of a right to control technological access to copyrighted work:

> In a similar vein, Chamberlain's proposed construction would allow any manufacturer of any product to add a single copyrighted sentence or software fragment to its product, wrap the copyrighted material in a trivial "encryption" scheme, and thereby gain the right to restrict consumers' rights to use its products in conjunction with competing products. In other words, Chamberlain's construction of the DMCA would allow virtually any company to attempt to leverage its sales into aftermarket monopolies – a practice that both the antitrust laws, and the doctrine of copyright misuse normally prohibit.[44]

Indeed, even if the grant to prohibit circumvention were a property right, it would still not serve to repeal by implication the rights of those who lawfully possess copyrighted property. If lawful ownership of encryption-protected property conferred two separate and distinct rights on the copyright holder, the latter would be empowered to prohibit lawful access to the copyrighted content. In the instant case, the owner of a Chamberlain GDO has a legal right, that is, a fair use right under copyright law, to use a universal transmitter manufactured by a third-party to access the copyrighted software to open the garage door:

> Chamberlain's proposed construction would allow copyright owners to prohibit *exclusively fair* uses even in the absence of any feared foul use. It would therefore allow any copyright owner, through a combination of contractual terms and technological measures, to repeal the fair use doctrine with respect to an individual copyrighted work – or even selected copies of that copyrighted work. Again, this implication contradicts § 1201(c)(1) directly. Copyright law itself authorizes the public to make certain uses of copyrighted materials. Consumers who purchase a product containing a copy of embedded software have the inherent legal right to use that copy of the software. What the law authorizes, Chamberlain cannot revoke.[45]

While the DMCA granted copyright holders additional legal protection, it did not rescind the basic bargain granting the public noninfringing and fair uses of copyrighted materials. The act also permits the various express beneficial uses of circumvention technology, such as those exempted under § 1201(d),(f),(g), and (j) from the anticircumvention ban.

The court held that 17 U.S.C. § 1201 prohibits only forms of access that bear a reasonable relationship to the protections that the Copyright Act otherwise affords owners. It speculated that its holding might "create some uncertainty and consume some judicial resources," but nevertheless concluded that it is the only meaningful reading of the statute.

Building upon its assumptions, the court articulated a very specific test to determine what a plaintiff alleging a violation of § 1201(a)(2) trafficking must establish:

> A plaintiff alleging a violation of § 1201(a)(2) must prove: (1) ownership of a valid *copyright* on a work, (2) effectively controlled by a *technological measure,* which has been circumvented, (3) that third parties can now *access* (4) *without authorization,* in a manner that (5) infringes or facilitates infringing a right *protected* by the Copyright Act, because of a product that (6) the defendant either (i) *designed or produced* primarily for circumvention; (ii) made available despite only *limited commercial significance* other than circumvention; or (iii) *marketed* for use in circumvention of the controlling technological measure. A plaintiff incapable of establishing any one of elements (1) through (5) will have failed to prove a prima

> facie case. A plaintiff capable of proving elements (1) through (5) need prove only one of (6)(i), (ii), or (iii) to shift the burden back to the defendant. At that point, the various affirmative defenses enumerated throughout § 1201 become relevant.[46]

A copyright owner seeking to impose liability on an accused trafficker must demonstrate that the trafficker's device enables either copyright infringement or a prohibited circumvention. Here, Chamberlain established no connection between unauthorized use of its copyrighted software and Skylink's accused transmitter. Hence, the court affirmed the district court's grant of summary judgment to Skylink.

Conclusion

The major challenges presented by both the *Lexmark* and *Chamberlain* cases are to define the relationships between and effects of the anticircumvention restrictions on reverse engineering, in particular, and underlying principles of copyright law, in general. These cases represent the application of anticircumvention principles to access control technology in a broader market of consumer goods. In both cases, a Court of Appeals read the restrictions on circumvention more narrowly than the interpretation asserted by the plaintiff, placing the initial burden on the plaintiff to establish not just the fact of circumvention, but its wrongfulness as well. In both cases, the courts strove to reconcile existing copyright law principles of fair use with the newer, independent constraints on circumvention. While these decisions may allay some of the DMCA's critics' concerns regarding anticircumvention and fair use, the facts before the courts ensures that they cannot define the outer limits for the use of access control technology as a means to assert control over durable goods.

References

[1] 17 U.S.C. § 1201 entitled "Circumvention of copyright protection systems."

[2] While a copyright owner's exclusive rights in a copyrighted work is set forth at 17 U.S.C. 106, exceptions to exclusive rights are set forth at, *e.g.,* 17 U.S.C. §§ 107, 108, 110, 111, 117.

[3] The statute was widely publicized in a successful suit by movie studios to prevent the posting over the Internet of a code to circumvent the Content Scrambling System, an encryption program for digital motion pictures. Universal City Studios v. Corley, 273 F.3d 429 (2d Cir. 2001).

[4] H.R. Rept. 105-551, Part 2, 105th Cong., 2d Sess. 24-25 (1998) quoting a letter from copyright law professors arguing that enactment of anticircumvention legislation "would represent an unprecedented departure into the zone of what might be called paracopyright – an uncharted new domain of legislative provisions designed to strengthen copyright protection by regulating conduct which traditionally has fallen outside the regulatory sphere of intellectual property law."

[5] 17 U.S.C. § 107.

[6] Sega Enterprises Ltd. v. Accolade, Inc., 977 F.2d 1510 (9th Cir. 1993); *See also* Sony Computer Entertainment v. Connectix Corp., 203 F.3d 596 (9th Cir. 2000); Bateman v.

Mnemonics, Inc., 79 F.3d 1532, 1539 (11th Cir. 1996); Atari Games Corp. v. Nintendo, 975 F.2d 832, 843 (Fed. Cir. 1992).

[7] *See* Julie E. Cohen, *Reverse Engineering and the Rise of Electronic Vigilantism: Intellectual Property Implications of "Lock-Out" Programs,* 68 S. CAL. L. REV. 1091 (1994-1995).

[8] 977 F.2d 1510 (9th Cir. 1993).

[9] *Id.* at 1523. (Citations omitted.)

[10] 17 U.S.C. § 1201(f)(1).

[11] *See* Bowers v. Baystate Technologies, Inc., 320 F.3d 1317 (Fed. Cir.), *cert. denied*, 539 U.S. 928 (2003)(Copyright Act did not preempt or narrow scope of a shrink wrap license agreement that prohibited reverse engineering.) *See also,* Davidson & Assoc., Inc. v. Internet Gateway, 334 F.Supp.2d 1164 (E.D.Mo. 2004).

[12] Assessment Technologies of WI v. Wiredata, Inc., 350 F.3d 640, 647 (7th Cir. 2003).

[13] *See* Dan L. Burk, *Anticircumvention Misuse,* 50 U.C.L.A. L. REV. 1095, 1135 (2003). The author advocates development of a judicial doctrine of "anticircumvention misuse," akin to copyright misuse, to curb anticompetitive applications of access-control technology. "[A] finding of misuse would be proper where the ends to which the anticircumvention right is put exceed the reasonable grant of the right. For this standard to have any definite structure it will be necessary to determine what the bounds of the anticircumvention grant might be."

[14] 253 F.Supp.2d 943 (E.D.Ky. 2003), *rev'd*, 387 F.3d 522 (6th Cir. 2004).

[15] *See also* Sony Computer Entertainment America. Inc. v. Gamemasters, 87 F.Supp.2d 976 (N.D.Ca. 1999)(no copyright misuse where plaintiff sued to prevent defendant from selling counterfeit accessories which circumvented access controls on plaintiff's game console and CD-ROMs.)

[16] 253 F.Supp.2d at 969.

[17] *Id.* at 967.

[18] Sections 1201(f)(2) and (3) provide that a person may develop a circumvention device and make that circumvention device available to others "*solely* for the purpose of enabling interoperability of *an independently created computer program* with other programs, and *to the extent that doing so does not constitute infringement under this title or violate applicable law other than this section.*" 17 U.S.C. § 1201(f)(3) (emphasis added).

[19] 387 F.3d 522 (6th Cir. 2004).

[20] 17 U.S.C. § 102(b).

[21] 387 F.3d at 535.

[22] *Id. Id.* at 543.

[23] *Id.* at 545. (Citation omitted; emphasis in original.)

[24] *Id.* at 546.

[25] *Id.* at 547.

[26] *Id.* at 548.

[27] *Id.* at 549.

[28] *Id.* at 551.

[29] Chamberlain Group, Inc. v. Skylink Technologies, Inc., 292 F.Supp.2d 1023 (N.D.Ill. 2003) (*Chamberlain I*).

[30] *Id.* at 1035.

[31] Chamberlain Group, Inc. v. Skylink Technologies, Inc., 292 F.Supp.2d 1040 (N.D.Ill. 2003) (*Chamberlain II*), *aff'd* 381 F.3d 1178 (Fed. Cir. 2004).

[32] 17 U.S.C. § 1201(a)(3)(A) provides that to circumvent a technological measure "means to descramble a scrambled work, to decrypt an encrypted work, or otherwise to avoid, bypass, remove, deactivate, or impair a technological measure, without the authority of the copyright owner[.]"

[33] 292 F.Supp.2d at 1044.

[34] *Id.* at 1045.

[35] *Id.* at 1046.

[36] 381 F.3d 1178, 1185, 1191 (Fed. Cir. 2004). The court distinguished the case before it from Universal City Studios v. Corley, 273 F.3d 429 (2d Cir. 2001), which focused on First Amendment issues in relation to the DMCA's anticircumvention provisions.

[37] 381 F.3d at 1187.

[38] *Id.* at 1192-93.

[39] *Id.* at 1193.

[40] "Circumvention is not a new form of infringement but rather a new violation prohibiting actions or products that facilitate infringement[.]" *Id.* at 1197.

[41] This interpretation, the court noted, helps explain "why Chamberlain's warranty conditions and website postings cannot render users of Skylink's Model 39 <unauthorized' users for the purposes of establishing trafficking liability under the DMCA." *Id.* at 1194.

[42] *Id.* at 1200.

[43] *Id.* at 1201. (Citations and footnote omitted.)

[44] *Id.* at 1202. (Footnote omitted)(Emphasis in original).

[45] *Id.* at 1203. (Emphasis in original).

In: New Intellectual Property Issues
Editor: John V. Luong, pp. 145-156

ISBN: 978-1-60021-211-6
© 2007 Nova Science Publishers, Inc.

Chapter 8

COPYRIGHT LAW: DIGITAL RIGHTS MANAGEMENT LEGISLATION*

Robin Jeweler

SUMMARY

Digital Rights Management (DRM) refers to the technology that copyright owners use to protect digital media. This report surveys several of the DRM bills that were introduced in the 107th Congress and those that are pending in the 108th Congress. Generally, the bills are directed at two separate goals. One goal is to increase access to digitally-protected media for lawful purposes. The other attempts to thwart digital piracy and would do so by enhancing civil and criminal sanctions for digital (and traditional) copyright infringement and educating the public about the rights of copyright holders.

Although no bills were enacted during the 107th Congress, two of the bills focusing on access have been reintroduced in the 108th Congress. Representatives Boucher and Lofgren reintroduced their bills from the 107th Congress. They are H.R. 107, the "Digital Media Consumers' Rights Act of 2003" and H.R. 1066, respectively. H.R. 1066 is renamed the "Benefit Authors without Limiting Advancement or Net Consumer Expectations (BALANCE) Act of 2003." And Senator Wyden introduced S. 692, a labeling disclosure bill entitled the "Digital Consumers Right to Know Act."

S. 1621, the "Consumers, Schools, and Libraries Digital Management Awareness Act of 2003," addresses several DRM issues. It would prohibit the Federal Communication Commission from establishing mandatory technology standards and require disclosure requirements for access controlled digital media and consumer electronics. It also addresses the subpoena process by which copyright owners acquire personal information about suspected infringers.

Bills addressing piracy include H.R. 4077, the "Piracy Deterrence and Education Act of 2004"; S. 1932, the "Artists Rights and Theft Prevention Act of 2004"; and, S. 2237, the "Protecting Intellectual Rights Against Theft and Appropriation Act of 2004." S. 1932 and S. 2237 have been reported favorably by the Senate Judiciary Committee. Because digital transmission poses the greatest distribution risk to entertainment content owners, these bills attempt to thwart the initial unauthorized copying and/or uploading to the Internet. Hence, sanctions for illegal distribution of pre-release commercial works and surreptitious recording of movies in theaters are emphasized.

---

* Excerpted from CRS Report RL32035, Updated May 28, 2004

BACKGROUND

Digital technology has radically altered the landscape of copyright law. The potential for unauthorized but near perfect replication of digital media poses new challenges to copyright owners. Copyright law gives a copyright holder the exclusive right to reproduce, adapt, distribute, perform publicly and display protected material for a limited term.[1] Historically, technology (or lack thereof) presented an obstacle to wide-spread piracy. For example, repeated copying of analog video or audio tapes could result in a degradation of the quality of the reproduction; packaging and transportation for distribution could be cumbersome. There are no comparable impediments to copying and distributing media in a digital format.

The legal basis for protecting copyright has traditionally been through the initiation of a civil proceeding by the copyright holder against the infringer for injunctive relief and/or money damages, although there are criminal sanctions for willful infringement as well.[2] The efficacy of this remedy is diminished in a digital environment where distribution may be decentralized, instantaneous, and global. Unauthorized peer-to-peer (P2P) file sharing of music illustrates this problem. Instituting a civil suit against thousands of individuals for *each* unauthorized download has traditionally been presumed to be infeasible. Recently, however, the Recording Industry Association of American (RIAA) has embarked on an aggressive litigation enforcement effort against college students and others who upload or download copyrighted sound recordings on P2P file-sharing sites.[3] The efficacy of this effort remains to be seen. Nevertheless, copyright owners believe that *prevention* of piracy is preferable. The technology-based approaches and mechanisms that copyright owners utilize to protect digital media are referred to as digital rights management (DRM).

Congress has enacted two laws to date which facilitate DRM to enhance copyright protection. The Audio Home Recording Act (AHRA) of 1992 effects a technology-based regulatory program for consumer goods designed to copy analog and digital musical recordings.[4] It requires manufacturers and distributors of audio recording devices to employ copy control technology. However, a "digital musical recording" is defined as a "material object" that does *not* include "one or more computer programs."[5] Hence, the AHRA does not cover songs fixed on computer hard drives and extends only to recordings from the material objects in which songs are otherwise normally fixed, such as recorded compact discs (CDs), digital audio tapes, audio cassettes, long-playing albums, digital compact cassettes, and mini-discs.[6]

The act requires consumer goods manufacturers to incorporate the Serial Copyright Management System (SCMS) into digital audio recording devices. SCMS is technology that sends, receives, and acts upon information about the generation and copyright status of the files that it plays. It allows copies to be made from an authorized recording, but prevents the SCMS-equipped machine from making copies of copies. The AHRA prohibits circumvention of the SCMS system as well. In consideration for permission to facilitate consumer copying of music recordings, manufacturers are required to pay music royalties based on sales of the devices. And, as a consequence of and in consideration for the technology-limited copying and royalty payments program, manufacturers, importers, and distributors of audio recording devices, and consumers who use them for noncommercial use, are protected from suit for copyright infringement.

A more recent – and more controversial – DRM law is the Digital Millennium Copyright Act (DMCA) of 1998. This law added a new chapter 12 to the Copyright Act entitled "Copyright Protection and Management Systems."[7] Subject to relatively narrow exceptions, this law makes it illegal to circumvent a technological copyright-control measure. This includes activity to descramble a scrambled work, to decrypt an encrypted work, or to avoid, bypass, remove, deactivate, or impair a technological measure, without the authority of the copyright owner.[8] In contrast to copyright infringement, which prohibits unauthorized or unexcused *use* of copyrighted material, the anti-circumvention provisions of the DMCA prohibit the design, manufacture, import, offer to the public, or trafficking in technology produced to circumvent copyright encryption programs, regardless of the actual existence or absence of copyright infringement. Even though the anticircumvention provisions of the DMCA have, to date, been upheld by the courts, critics argue that they have a chilling effect on rights of free speech and that their implementation will thwart the public's ability to access copyrighted works, which is ultimately necessary in order to exercise "fair use."[9]

Fair Use

The doctrine of "fair use" is a limitation upon a copyright holder's exclusive rights. It permits the public to use a copyrighted work for limited purposes, such as criticism, comment, news reporting, teaching, scholarship or research.[10] And, although the concept of "personal use" (i.e., copying lawfully acquired copyrighted materials for one's personal use) is not expressly protected by statute, it is widely-accepted and judicially sanctioned.[11] Fair use protects the public interest in a free exchange of ideas and discourse.

The ever-changing state of technology and DRM laws raise many issues, several of which paradoxically confound one another. To the extent that copyrighted digital material is not encrypted, it may be subject to piracy on a massive scale. But, as digital material is increasingly encrypted to protect against piracy, the public is in jeopardy of restricted access, which may impede the exercise of fair use. And, some observers assert that protection controls give copyright holders more exclusive control over their creations than the copyright law intends.[12] Content owners, however, argue that allowing *limited* circumvention only to facilitate fair use, including personal use, is impracticable; once the circumvention technology becomes publicly available, its protective value is compromised.

Members of 107th Congress responded by introducing bills which address two sides of the issue – piracy prevention and fair use access. Legislative proposals would have mandated government-sponsored encryption technology and enhanced content owners' abilities to fight P2P piracy over the Internet. Other bills were intended to clarify and expand content users' fair use, including personal use, access to digital media. While none of these bills were enacted during the 107th Congress, the underlying policy issues have not been resolved, and the matter continues to be the subject of interest in the 108th Congress. This report surveys several of the DRM bills introduced in the 107th and 108th Congresses.

LEGISLATION IN THE 107TH CONGRESS

Bills Promoting Enhanced DRM Anti-piracy Protection in the 107th Congress

S. 2048, 107th Cong., 2d Sess. (2002), the "Consumer Broadband and Digital Television Promotion Act". This bill, introduced on March 21, 2002 by Senator Hollings, would have directed digital media device manufacturers, consumer groups, and copyright owners to attempt to reach an agreement on security system standards for use in digital media devices and encoding rules within a year after enactment. If parties were unable to agree on acceptable standards, the Federal Communications Commission (FCC), in consultation with the Copyright Office, would do so. With respect to prospective encoding rules, the bill stipulated that the rules "shall take into account the limitations on the exclusive rights of copyright owners, including the fair use doctrine." It further provided that "[n]o person may apply a security measure that uses a standard security technology to prevent a lawful recipient from making a personal copy for lawful use... ."[13]

The bill would have implemented the standards by requiring interactive computer services to incorporate security measures associated with standard security technologies and by requiring manufacturers, importers and sellers of digital media devices to include the security technologies. It would have prohibited removal or alteration of the technology from the devices. How fair use access and the security technology would interface was not expressly addressed.

H.R. 5211, 107th Cong., 2d Sess. (2002), a bill "to limit the liability of copyright owners for protecting their works on peer-to-peer networks". Sponsored by Representative Berman, this bill was designed to create "a safe harbor from liability so that copyright owners could use technological means to prevent the unauthorized distribution of that owner's copyrighted works via a P2P network."[14] It would have added a new section to the Copyright Act, 17 U.S.C. § 514, which would exempt copyright owners from liability under state and federal law for

> disabling, interfering with, blocking, diverting, or otherwise impairing the unauthorized distribution, display, performance, or reproduction of his or her copyrighted work on a publicly accessible peer-to-peer file trading network, if such impairment does not, without authorization, alter, delete, or otherwise impair the integrity of any computer file or data residing on the computer of a file trader.[15]

The bill included exceptions to the safe harbor for copyright owners, and required them to notify the Department of Justice before employing specific blocking technologies. It created a new cause of action, in addition to existing ones, for file traders to deter harassment or abuse of P2P networks by copyright owners.

Because the bill aimed to allow the prevention of unauthorized file trading over decentralized P2P networks, its sponsors suggested that it would not adversely impact consumers' fair use of digital media.[16]

Bills Addressing Digital Access

In addition to the bills noted below, Senator Wyden and Representative Cox introduced, respectively, Senate and House joint resolutions entitled the "Consumer Technology Bill of Rights."[17] They were reportedly based upon a proposal of the same name by the advocacy group, DigitalConsumer.org.[18] Their premise was that copyright law should not curtail consumers' fair use rights with respect to digital and electronic entertainment media. The resolutions' enumeration of consumer rights included the right to use technology for

- "time-shifting" (i.e., recording legally acquired audio or video for later listening or viewing);
- "space-shifting" (i.e., using legally acquired content in different places);
- making backup or archival copies;
- using legally acquired content on the electronic platform or device of choice; and,
- translating legally acquired content into comparable formats.

Discussed below are two bills addressing digital fair use which were introduced in the latter part of the 107th Congress.

H.R. 5522, 107th Cong.,2d Sess. (2002), the "Digital Choice and Freedom Act". On October 2, 2002, Representative Zoe Lofgren introduced this bill. Among its findings is the observation that "[D]igital technology threatens the rights of copyright holders. Perfect digital copies of songs and movies can be publicly transmitted without authorization to thousands of people at little or no cost. On the other hand, technological control measures give copyright holders the capacity to limit non-public performances and threaten society's interest in the free flow of ideas, information and commerce."[19]

In order to recalibrate the balance between the copyright interests of authors and society, H.R. 5522 would have amended the Copyright Act to effect three goals:

- To expressly provide that it is not a copyright infringement for a person who lawfully possesses or receives a transmission of a digital work to reproduce, store, adapt or access it for archival purposes or to transfer it to a preferred digital media device in order to effect a non-public performance or display;
- To amend 17 U.S.C. § 109,[20] to allow one who lawfully possesses a digital work to sell or otherwise dispose of it by means of a transmission to a single recipient, provided that the owner does not retain his or her copy; and
- To amend the DMCA, 17 U.S.C. § 1201, to permit circumvention of copyright encryption technology, including the manufacture and import of, and trafficking in technology, if it is necessary to enable a non-infringing use and the copyright owner fails to make available the necessary means for circumvention.

H.R. 5544, 107th Cong., 2d Sess. (2002), the "Digital Media Consumers' Rights Act of 2002". Introduced on October 4, 2002 by Representatives Boucher and Doolittle, the Digital Media Consumers' Rights Act addresses copy-protected (*i.e*, non-standard) audio CDs through consumer disclosure via labeling requirements. Specifically, the bill would have amended the Federal Trade Commission Act by adding a new section entitled "Inadequately

Labeled Copy-Protected Compact Discs." The new labeling requirements were intended to notify consumers when a non-standard CD has copy-protection measures which could preclude playing on and/or copying to a computer hard drive or other consumer electronic devices. The Federal Trade Commission would be empowered to engage in rulemaking regarding audio CD labeling to prevent consumer confusion about playability and recordability.

The bill, in a vein similar to H.R. 5522, would have amended the DMCA to broaden the exemption for scientific research from the anti-circumvention rule; to permit circumvention for non-infringing uses; and, to permit the manufacture and sale of circumvention software capable of a significant non-infringing use.

BILLS INTRODUCED IN THE 108TH CONGRESS

Bills Addressing Digital Access and Disclosure

Representatives Boucher and Lofgren have reintroduced their bills from the 107th Congress. In the 108th Congress, Representative Boucher's bill is H.R. 107, the "Digital Medial Consumers Rights Act of 2003." The House Subcommittee on Commerce, Trade, and Consumer Protection held a hearing on H.R. 107 on May 12, 2004. Representative Lofgren's bill, H.R. 1066, is renamed the "Benefit Authors without Limiting Advancement or Net Consumer Expectations (BALANCE) Act of 2003."

S. 692, 108th Cong., 1st Sess. (2003), the "Digital Consumer Right to Know Act." Introduced by Senator Wyden, this bill emphasizes and would require disclosure of DRM anti-piracy protections that would restrict consumers' use of digital content. Specifically, the Federal Trade Commission (FTC) is directed to issue rules governing disclosures of technological features that limit the practical ability of a purchaser "to play, copy, transmit, or transfer such content on, to, or between devices or classes of devices that consumers commonly use ... prior to sale."[21] Examples of limitations subject to the disclosure requirement include:

- limitations on "time shifting" (recording for later viewing or listening) of free over-the-air and certain subscription packaged audio or video programming;
- limitations on reasonable and noncommercial use of legally acquired video or audio programing to facilitate "space shifting" (recording for use in different physical locations), including the transfer of the content to different electronic platforms or devices;
- limitations on making backup copies of legally acquired content;
- limitations on using limited excerpts of legally acquired content for criticism, comment, news reporting, teaching, scholarship, or research; and
- limitations on engaging in the transfer or sale of legally acquired content.

The goal of the legislation is to address legitimate consumer expectations regarding how they may use and manipulate content in concert with developing technology, and to promote development of an acceptable balance between protecting against piracy and preserving utility and flexibility for consumers.[22]

S. 1621, 108th Cong., 1st Sess. (2003), the "Consumers, Schools, and Libraries Digital Rights Management Awareness Act of 2003". Introduced by Senator Brownback on September 16, 2003, the DRM Awareness Act addresses several issues, including mandatory technology standards, disclosure requirements for access controlled digital media and consumer electronics, the subpoena process by which copyright owners acquire personal information about suspected infringers, and the secondary market for digital consumer goods. Among the bill's findings are:

- it is not in public interest for the federal government to mandate the inclusion of access or redistribution control technologies used with digital media into consumer electronics;
- access controlled compact discs have created confusion and inconvenience for consumers, educational institutions, and libraries;
- it is not in the public interest for Internet service providers to disclose personal information about subscribers for whom they transport electronic communications; and
- it is not in the public interest to allow access or redistribution control technologies to limit the secondary market for digital media products.

To this end, S. 1621 would *prohibit* the Federal Communications Commission from establishing mandatory access control technology (including redistribution control technology) standards for consumer digital media machines and devices. The FCC could not require manufacturers and importers of digital media devices to incorporate access or redistribution control technology. It would, however, grandfather in two pending FCC rulemaking proceedings, namely, "cable plug and play" and digital broadcast copy protection, subject to the requirement that objective standards, not specific technologies, be adopted.

In addition, the bill would:

- direct the Federal Trade Commission to establish an advisory committee to study and report on the ways in which access and redistribution control technology affect consumers, educational institutions, and libraries and how to better inform them of the impact of such technologies;
- direct the FTC to establish labeling requirements to inform consumers of the existence of access or redistribution control technology-protected digital products;
- address litigation over the DMCA's requirement that Internet service providers (ISPs) respond to subpoenas obtained by copyright holders who suspect copyright infringement. The bill provides that ISPs *not* make information about its subscribers available unless a subpoena is issued pursuant to state law or the Federal Rules of Civil Procedure, or unless the information requested relates to allegedly infringing digital products residing on the system or network of the ISP;
- create a consumer "first sale" doctrine for digital media by providing that an owner may transmit a digital product to a single recipient as long as the transmission technology contemporaneously deletes the transmitter's copy; and
- prohibit manufacturers and vendors of digital media from imposing any access or redistribution control technology that prevents a consumer from donating the item to

an educational institution or library, or that limits consumer resale or donations to specific venues or distribution channels.

Bills Addressing Copyright Piracy

H.R. 2517, 108th Cong., 1st Sess. (2003), the "Piracy Deterrence and Education Act of 2003". Introduced on June 19, 2003 by Representative Lamar Smith, this bill takes a two-pronged approach to copyright piracy. It would enhance criminal copyright infringement enforcement and public education about use of copyrighted material. Section 1 of the bill sets forth a lengthy recitation of congressional findings of problems that warrant corrective legislation.[23] Among the findings are:

- IP theft through electronic means causes great economic damage;
- Many computer users do not know that copyright laws apply to the Internet or believe that they will not be caught or prosecuted for their conduct;
- Use of P2P systems may pose serious security and privacy threats to computer users;
- It is important that federal law enforcement agencies prosecute theft of copyright and that the public be educated about the security and privacy risks associated with being connected to unauthorized P2P networks; and
- Formal copyright registration requirements unnecessarily burden criminal and civil litigation efforts to enforce the laws protecting copyright.

The bill directs the Federal Bureau of Investigations (FBI) to develop programs to deter copyright infringement over the Internet, including the issuance of appropriate warnings and facilitating information sharing about infringing activities among law enforcement agencies, ISPs and copyright owners. The Attorney General is directed to designate at least one agent to investigate IP theft in any unit of the Department of Justice (DOJ) responsible for investigating computer hacking and IP crimes.

The DOJ is directed to establish an "Internet Use Education Program" to educate the general public about the damage resulting from IP theft; the privacy and security risks of P2P file-sharing to obtain unauthorized copies of copyrighted work; and, to coordinate and consult with the Departments of Education and Commerce regarding copyright law and Internet use. The Attorney General will also establish criteria for use by specified copyright owners of the seal of the FBI for deterrent purposes in connection with digital works of authorship.

The bill also waives certain copyright registration requirements, considered to be formalities, which hinder or delay enforcement actions by the government, including actions to prevent importation of infringing materials by the Bureau of Customs and Border Protection of the Dept. of Homeland Security.

H.R. 2752, 108th Cong., 1st Sess. (2003), the "Author, Consumer, and Computer Owner Protection and Security Act (ACCOPS) Act of 2003." Introduced by Representative Conyers as a companion bill to H.R. 2517, ACCOPS would increase international and domestic anti-piracy IP law enforcement efforts. In addition to increasing appropriations for criminal law enforcement and procedures directing U.S. cooperation with foreign authorities in international investigations, the bill increases criminal sanctions for domestic copyright infringement. Title III of the bill would, among other things:

- amend 17 U.S.C. § 506 to provide that willful, unauthorized uploading of a single copyrighted work on the Internet satisfies the standards for a felony as opposed to a misdemeanor offense.
- require P2P file-swapping software distributors to give notice of potential security risks posed by the software and to receive consent from the downloader of such software;
- make it a federal crime to surreptitiously record a movie being performed in a movie theater; and
- direct the courts to consider that providing misleading or false contact information to a domain registry by a domain name registrant is evidence of "willfulness" with respect to any copyright infringement committed through the use of the domain name.

H.R. 4077, 108th Cong., 2d Sess. (2004), the "Piracy Deterrence and Education Act of 2004." Introduced by Representative Smith, this bill combines aspects of H.R. 2517 and H.R. 2752, discussed *supra*. The bill was introduced, marked up, and reported by the House Subcomm. on Courts, the Internet, and Intellectual Property on March 31, 2004.

Section 2 of the bill makes findings of fact similar to those set forth in H.R. 2517. Section 3 directs the FBI, in consultation with the Copyright Office, to develop programs to deter the public from copyright infringement over the Internet, including the issuance of appropriate warnings, and programs to facilitate information sharing about infringing activities among law enforcement agencies, ISPs and copyright owners. The Attorney General, in section 4, is directed to designate at least one agent to investigate IP theft in any unit of the DOJ responsible for investigating computer hacking and IP crimes.

The DOJ is directed to establish an "Internet Use Education Program" to educate the general public about the damage resulting from IP theft; the privacy and security risks of P2P file-sharing to obtain unauthorized copies of copyrighted work; and, to coordinate and consult with the Copyright Office and the Department of Commerce regarding copyright law and Internet use. The Program will develop sector-specific materials for Internet users where criminal copyright infringement is a concern. Section 5.

Section 6 would amend 17 U.S.C. § 411 to permit the government to file copyright infringement actions prior to copyright registration. Registration is ordinarily a prerequisite to filing an action for infringement. Section 7 sets forth authorized appropriations.

Section 8 is entitled the "Artists Rights and Theft Prevention Act of 2004 or the ART Act." (*Cf.* S. 1932 *infra*). The ART Act would add a new law, 18 U.S.C. § 2319B, expressly prohibiting unauthorized recording of motion pictures in a motion picture exhibition facility. Offenders would be subject to imprisonment for three to six years and forfeiture or destruction of the bootlegged copies. It would permit a victim of the crime to submit a victim impact statement to a probation officer.

Section 9 sets forth a Sense of the Congress on the need to take steps to prevent illegal activity on P2P services. It lists many findings regarding the widespread use of P2P file-sharing technology, including massive volumes of illegal activity such as distribution of child pornography, viruses, and confidential personal information. It identifies studies documenting the ways in which children are exposed to pornography through P2P technology. It concludes with a Sense of Congress that while responsible software developers should be recognized and commended, Congress and the executive branch should consider all appropriate measure to protect consumers and children and prevent illegal activity.

Section 10, entitled "Enhancement of Criminal Copyright Infringement," would amend 17 U.S.C. § 506 dealing with criminal offenses. Activity constituting criminal infringement would be expanded. Current § 506(a) makes criminal willful infringement for purposes of commercial advantage or private gain, or willful reproduction or distribution within 180 days of 1 or more copyrighted works with a retail value of more than $1000. New categories are: within any 180-day period, knowingly, with reckless disregard of the risk of further infringement, distributing by electronic means or otherwise (1) 1000 or more copies of 1 or more copyrighted works; (2) 1 or more copies of 1 work with a total retail value of more than $1000; or, (3) 1 or more copies of 1 or more copyrighted pre-release works. Evidence of reproduction alone will not be sufficient to establish criminal intent. Punishment may include fines and/or imprisonment for between 3 and 10 years as required by 18 U.S.C. § 2319, as amended by the bill. Under current § 2319, imprisonment for not more than five years is authorized only for the distribution "of at least 10 copies or phonorecords, of 1 or more copyrighted works, which have a total retail value of more than $2500[.]"

17 U.S.C. 504 would be amended to permit copyright owners to seek civil damages for infringement of pre-release works.

Section 11 directs the U.S. Sentencing Commission to review and, if appropriate, amend sentencing guidelines for persons convicted of IP crimes. Emphasis is on ensuring that guidelines are sufficiently stringent to deter and punish such offenses, including those involving pre-release copyrighted work. The Commission must also determine wether the definition of "uploading" in its Guidelines is adequate to address losses attributable to unauthorized distribution of copyrighted work over the Internet.

S. 1932, 108th Cong., 1st Sess. (2003), the "Artists Rights and Theft Prevention Act of 2004." Introduced on November 22, 2003 by Senator Cornyn, the bill was reported favorably with an amendment in the nature of a substitute by the Senate Judiciary Committee on April 29, 2004. If enacted, S. 1932, the "ART Act," would add new criminal penalties for unauthorized recording or filming of motion pictures in a theater. It is intended to stem bootlegging and unauthorized distribution of "*pre-release* commercial works."

Movie studios have complained that all too frequently an unauthorized version of a film is available online even before it is commercially released. Problems have been attributed to piracy by people in the film industry who have access to pre-release commercial works.[24] Specifically, S. 1932 would add a new law, 18 U.S.C. § 2319B, expressly prohibiting unauthorized recording of motion pictures in a motion picture exhibition facility.

The provision is conceptually related to current 18 U.S.C. § 2319A which establishes criminal sanctions for unauthorized filming or recording of live musical concerts.[25] S. 1932 would subject offenders to imprisonment for three to six years and forfeiture or destruction of the bootlegged copies. Movie theaters and exhibitors receive civil and criminal immunity from liability for a reasonable detention for questioning or arrest of any person suspected of violating the law. It would permit a victim of the crime to submit a victim impact statement to a probation officer.

S. 1932 would establish another category of criminal infringement: unauthorized distribution of a pre-release commercial copyrighted work.[26] Section 4 of the bill adds a new class of prohibited activity to 17 U.S.C. § 506 governing criminal copyright infringement. § 506(a) currently defines criminal infringement as willfully infringing for (1) commercial advantage or private financial gain or (2) by reproducing or distributing within 180 days one or more copyrighted works having a retail value of $1000. A new category – knowingly making a work being prepared for commercial distribution available on a computer network accessible to the public – would be added. Works covered include computer programs, motion pictures, and sound recordings. Punishment includes fines and/or imprisonment for 3 to 10 years.

The 5 of the bill directs the Copyright Office to establish procedures to allow preregistration of a work that is being prepared for commercial distribution and has not been published.[27] The work must be of a class that the Register determines suffers a history of pre-commercial distribution infringement. Copyright registration facilitates an action for infringement.

Section 6 directs the U.S. Sentencing Commission to review, and if appropriate, amend the federal sentencing guidelines and policy statements applicable to persons convicted of IP crimes.

S. 2237, 108th Cong., 2d Sess. (2004), the "Protecting Intellectual Rights Against Theft and Appropriation Act of 2004." Introduced on March 25, 2004 by Senators Leahy and Hatch, the PIRATE Act was reported by the Senate Judiciary Committee on April 29, 2004. The bill would add a new 17 U.S.C. § 506a to the Copyright Act entitled "Civil penalties for violations of section 506" which authorizes the U.S. Attorney General to file civil copyright infringement actions in a U.S. district court against any one who meets the standards for criminal infringement under § 506. Upon proof of conduct by a preponderance of the evidence, a defendant would be subject to a civil penalty under § 504, which encompasses actual damages, profits, and statutory damages. Although imposition of a civil penalty would not preclude additional civil or criminal actions, any restitution received by a copyright owner as a result of a civil action brought by the Department of Justice would be offset against an award of damages in a subsequent suit brought by the copyright owner.

The bill authorizes funding for a civil copyright enforcement litigation training and pilot programs for personnel in the DOJ and U.S. Attorneys Offices. The DOJ is directed to report annually on litigation under a newly enacted § 506a.

REFERENCES

[1] 17 U.S.C. § 106.

[2] *Id.* at §§ 501- 505.

[3] See, Frank Ahrens, *4 Students Sued Over Music Sites; Industry Group Targets File Sharing at Colleges,* The Washington Post, April 4, 2003 at E1; Ted Bridis, *RIAA's Subpoena Onslaught Aimed at Illegal File Sharing,* THE WASHINGTON POST, July 19, 2003 at E1.

[4] 17 U.S.C. §§ 1001 - 1010.

[5] 17 U.S.C. § 1001(5).

[6] See, Recording Industry Ass'n of America v. Diamond Multimedia Sys., Inc., 180 F.3d 1073 (9th Cir. 1999).

[7] 17 U.S.C. § 1201 *et seq.*

[8] *Id.* at § 1201(a)(3)(A).

[9] For more detail, see CRS Report RL31827, *"Digital Rights" and Fair Use in Copyright Law* by Robin Jeweler (March 24, 2003).

[10] 17 U.S.C. § 107.

[11] See Sony Corp. v. Universal Studios, Inc., 464 U.S. 417 (1984)(authorizing consumer use of home videocassette recorders to "time-shift" television broadcasts) .

[12] See, e.g., Glynn S. Lunney, Jr., *The Death of Copyright: Digital Technology, Private Copying, and the Digital Millennium Copyright Act,* 87 VA. L. REV. 813 (2001).

[13] S. 2048, § 3(e).

[14] Introductory statement of Rep. Berman, 148 CONG. REC. E1395 (daily ed. July 25, 2002).

[15] H.R. 5211 at § 514(a).

[16] Note 14, *supra.* ("Because its scope is limited to unauthorized distribution, display, performance or reproduction of copyrighted works on publicly accessible P2P systems, the legislation only authorizes self-help measures taken to deal with clear copyright infringements. Thus, the legislation does not authorize any interdiction actions to stop fair or authorized uses of copyrighted works ... or any interdiction of public domain works.")

[17] S.J.Res. 51, 107th Cong., 2d Sess. (2002) and H.J.Res. 116, 107th Cong. 2d Sess. (2002).

[18] *Wyden Offers Digital Fair Use Resolution,* 64 BNA Patent, Trademark & Copyright J. 585 (Oct. 25, 2002).

[19] H.R. 5522 at § 2(5).

[20] 17 U.S.C. § 109. This provision, known as the "first sale" doctrine, permits the owner of a copyrighted book or record to sell or otherwise dispose of it without violating copyright holder's right to control distribution.

[21] S. 6692, § 3(b).

[22] 149 CONG. REC. S4327 (daily ed. March 24, 2003)(Statement of Sen. Wyden).

[23] See also, *The Piracy Deterrence and Education Act of 2003: Hearing on H.R. 2517 before the House Subcomm. on Courts, the Internet, and Intellectual Property,* 108th Congress, 1st Sess. (2003).

[24] See, e.g., Bernard Weinraub, *Advance Film Copies Halted for Oscar Voters,* N.Y. TIMES, Oct. 1, 2003 at [http://www.nytimes.com/2003/10/01/movies/01OSCA.html].

[25] See United States v. Moghadam, 175 F.3d 1269 (11th Cir. 1999), *cert. den.* 529 U.S. 1036 (2000) upholding 18 U.S.C. § 2319A under congressional authority to legislate pursuant to the Commerce Clause.

[26] 18 U.S.C. § 2319 sets forth conditions and penalties for criminal copyright infringement.

[27] Section 5 of the bill would amend 17 U.S.C. § 408.

In: New Intellectual Property Issues
Editor: John V. Luong, pp. 157-163

ISBN: 978-1-60021-211-6
© 2007 Nova Science Publishers, Inc.

Chapter 9

COPYRIGHT PROTECTION OF DIGITAL TELEVISION: THE "BROADCAST FLAG"[1]

Angie A. Welborn

SUMMARY

This report addresses the adoption of a "broadcast flag" system by the Federal Communications Commission (FCC) to protect digital television (DTV) broadcasts from unauthorized redistribution. The report also addresses the recent decision of the United States Court of Appeals for the District of Columbia Circuit reversing and vacating the FCC's broadcast flag report and order.

INTRODUCTION

Technological advances, a looming statutory deadline, and the need to reclaim analog spectrum occupied by television broadcasters has put digital television (DTV) on a fast track. At the same time, development of digital television has necessitated balancing competing interests – those of content holders, and those of the consumer and technological industries. Reconciling these interests has led to the development of a "broadcast flag" to combat unauthorized redistribution of content broadcast through digital television signals. The move to protect digital content has been given urgency by the Federal Communications Commission's (FCC) determination that broadcast transmissions be digital by December 31, 2006.[1]

WHAT IS DTV?

Digital Television (DTV) is a new television service representing the most significant development in television technology since the advent of color television in the 1950s. Three

---

[1] Excerpted from CRS Report RS22106, Updated May 11, 2005
This is an update of a report prepared by Carey Lening, Law Clerk with the American Law Division

major components of DTV service must be present for consumers to enjoy a fully realized "high definition" television viewing experience. First, digital programming must be available. Digital programming is content that is assembled with digital cameras and other digital production equipment. Second, digital programming must be delivered to the consumer via a digital signal. Third, consumers must have digital television equipment capable of receiving the digital signal and displaying digital programming for viewing.[2]

Developing a protocol for transmitting and receiving digital television in a way that accommodated competing interests has proved challenging. Digital content, like other media, can be relatively easily duplicated and distributed, especially with the aid of the Internet.[3] Unlike other types of content, duplication of digital information does not degrade the original. Whereas the quality of a VHS tape degrades after successive copies, a DVD may be copied almost infinitely with no effect on the quality of the medium. It is due to the ease and inexhaustible potential of copying digital media, coupled with the proliferation of peer-to-peer services (such as Morpheus and Kazaa), that content providers have greeted this new technology with some trepidation.

The Broadcast Flag

The "broadcast flag" is a combination of technical specifications and federal regulations designed to combat unauthorized redistribution of content broadcast through digital television (DTV) signals. Its adoption was prompted largely by the Federal Communications Commission's (FCC) determination that broadcast transmissions be digital by December 31, 2006.[4] The FCC imposed a transition to DTV in part to capitalize on the sharper images, CD-quality sound and wider screen angles that are available from advanced digital technologies. But in addition to the technological impetus, the FCC has also been motivated by the knowledge that broadcasters, upon receiving digital spectrum allotments, must relinquish their analog spectrum allotments to the FCC. The analog spectrum will in turn be auctioned for other commercial and public interests. Content providers, fearing widespread piracy that would endanger aftermarket sales (such as cable re-broadcast, and DVD sales), urged the FCC to provide for a means to protect their assets. Meanwhile, consumer electronics and information technologists, as well as consumer rights groups, came together in an effort to minimize the possible negative outcome that a wide-scale regulation might have imposed.

The technical specifications behind the broadcast flag were a compromise measure, premised on an understanding that more restrictive approaches (such as encrypted signals created at the source of transmission) imposed economically or technologically infeasible conditions. The compromise came after a consortium of content providers, consumer electronics and information technology groups came together, forming the Broadcast Protection Discussion Group (BPDG).[5] The result of this consortium was a "Final Report" published in June 2002, which was delivered to Representative Billy Tauzin, then-Chairman of the House Committee on Energy and Commerce. The report suggested a set of "robustness and compliance" rules for devices capable of demodulating digital television signals, which would require that such devices protect "flagged" content from being recorded by unauthorized devices. However, the flag itself would not require that all machines recognize it, and would act only as a means to halt unauthorized use in machines capable of detecting it.

In November 2003, the FCC published a Report and Order in the matter of digital broadcast content protection, which required all digital devices capable of receiving digital broadcast over-the-air, and sold after July 1, 2005, to incorporate a standard content-protection technology that would recognize the broadcast flag, and limit redistribution when the flag is recognized.[6] The FCC's regulations apply the flag mark to all devices and receivers that are capable of receiving digital content. Such devices include, but are not limited to: televisions, computers, digital video-recorders (such as TiVo), and DVD players. The broadcast flag itself is optional for broadcasters, allowing them to determine how much copy-protection they wish to impose on their digital broadcast content.[7]

Because the flag does not prevent the distribution of content to non-compliant devices, a consumer who continues to use an older television set (or theoretically, a noncompliant demodulator) will still be able to receive and copy television programs in non-digital form. In addition, digital television sets made prior to July 1, 2005, will still enjoy digital content with no obstruction. In citing its support for a flag-based approach over encryption or other means, the FCC noted concerns over "the implementation costs and delays" associated with other solutions.[8]

In addition to the "compliance" requirements imposed on receiving devices, the FCC also imposed a "robustness" requirement that forces makers of consumer devices to ensure that circumvention is difficult. The standard of care adopted by the FCC was that of "an ordinary user using generally available tools or equipment."[9]

FCC Authority

The FCC derives its regulatory authority over digital television from both direct and ancillary statutory authority.

Digital Television Implementation Under the Telecommunications Act of 1996

The Telecommunications Act of 1996 directed the FCC to promulgate regulations regarding the licensing of advanced television services. The act defined "advanced television services" as "television services provided using digital or other advanced technology."[10] In prescribing such regulations, the Commission was authorized to adopt such "technical and other requirements as may be necessary or appropriate to assure the quality of the signal used to provide advanced television services ... and prescribe such other regulations as may be necessary for the protection of the public interest, convenience, and necessity."[11]

Pursuant to the Telecommunications Act of 1996, the FCC has issued regulations regarding spectrum allocation for digital television stations and has established a timeline for the implementation of digital broadcasting by licensees.[12] At least one court has agreed that in regard to television digital tuners, the FCC possessed reasonable authority to act, based on an "unambiguous command of an act of Congress."[13]

Copyright Protection

While copyright protection generally lies outside the scope of the FCC, the Commission may exercise jurisdiction over matters not explicitly provided for by statute if the exercise is "reasonably ancillary to the effective performance of the Commission's various responsibilities for the regulation of television broadcasting."[14] The FCC has asserted that television receivers generally, and digital television receivers specifically, fall within the scope of that authority.[15] Under the FCC Report and Order, "pursuant to the doctrine of ancillary jurisdiction, we adopt use of the ... flag as currently defined for redistribution control purposes and establish compliance and robustness rules for devices with demodulators to ensure that they respond and give effect to the ... flag." However, the FCC initially put off deciding on permanent mechanisms for approving "content protection and recording technologies to be used in conjunction with device outputs." Instead, in its Report and Order, the FCC proposed examination of such issues at a later time, and established an interim certification process for currently-proposed devices.[16] In addition to the need to regulate television broadcasting, the FCC's action arguably protects broadcasters from any unreasonable loss in advertising revenue that may result from unauthorized sharing of copyrighted digital television broadcasts.

However, the FCC was careful to note that the "scope of our decision does not reach existing copyright law," and that its rulemaking established a "technical protection measure" that did not change the underlying "rights and remedies available to copyright holders." In addition, "this decision is not intended to alter the defenses and penalties applicable in cases of copyright infringement, circumvention, or other applicable laws."[17]

Possible Implications of the Broadcast Flag

While the broadcast flag is intended to "prevent the indiscriminate redistribution of [digital broadcast] content over the Internet or through similar means," the goal of the flag was not to impede a consumer's ability to copy or use content lawfully in the home, nor was the policy intended to "foreclose use of the Internet to send digital broadcast content where it can be adequately protected from indiscriminate redistribution."[18] However, current technological limitations have the potential to hinder some activities which might normally be considered "fair use" under existing copyright law.[19] For example, a consumer who wished to record a program to watch at a later time, or at a different location (time-shifting, and space-shifting, respectively), might be prevented when otherwise approved technologies do not allow for such activities, or do not integrate well with one another, or with older, "legacy" devices. In addition, future fair or reasonable uses may be precluded by these limitations. For example, a student would be unable to email herself a copy of a project with digital video content because no current secure system exists for email transmission.

In addition, some consumer electronics and information technology groups contend that the licensing terms for approving new compliant devices are limiting, and may potentially stifle innovation, especially with regard to computer hardware.[20] While the FCC in its Report and Order declined to establish formal guidelines for which "objective criteria should be used to evaluate new content protection and recording technology," it has stated an intention to take up these issues in the future.[21]

Finally, consumer rights and civil liberties groups worry about the possibility that such content protections will limit the free flow of information and hamper the First Amendment. This concern is expressed most prominently regarding news or public interest-based content, or works that have already entered the public domain. Despite suggestions raised by consumer rights groups, the FCC has so far declined to adopt language to prevent content providers from using the broadcast flag on such programs, largely because of the "practical and legal difficulties of determining which types of broadcast content merit protection from indiscriminate redistribution and which do not."[22]

LEGAL CHALLENGES TO THE BROADCAST FLAG

In October of 2004, the American Library Association (ALA), Association of Research Libraries, American Association of Law Libraries (AALL), Medical Libraries Association and others, petitioned the U.S. Court of Appeals for the District of Columbia Circuit to review the FCC's Report and Order. Bringing a challenge on behalf of "libraries, librarians and educators ... and ... television viewers and computer users," the petitioners, as parties to the agency proceedings, questioned the FCC's statutory authority to establish the broadcast flag system under the Communications Act of 1934. On May 6, 2005, the United States Court of Appeals for the District of Columbia Circuit granted the ALA's petition for review and reversed and vacated the Commission's order requiring DTV reception equipment to be manufactured with the capability to prevent unauthorized redistributions of digital content.[23]

In *American Library Association v. Federal Communications Commission*, the court of appeals determined that the FCC lacked the authority "to regulate apparatus that can receive television broadcasts when those apparatus are not engaged in the process of receiving a broadcast transmission."[24] The court noted that in adopting the broadcast flag rules, the Commission "cited no specific statutory provision giving [it] authority to regulate consumers' use of television receiver apparatus after the completion of the broadcast transmission."[25] The Commission's reliance on its ancillary jurisdiction under Title I of the Communications Act of 1934 was rejected by the court. The court found that while the jurisdictional grant under Title I plainly encompasses the regulation of apparatus that can receive television broadcast content, the Commission's regulatory authority does not extend beyond the actual receipt of such content by the apparatus in question. The court's decision was limited to the resolution of the question of whether the Commission had the authority to impose the broadcast flag requirements, and did not address the imposition of the broadcast flag requirements in terms of copyright law.

REFERENCES

[1] Federal Communications Commission, *In the Matter of Advanced Television Systems and Their Impact Upon the Existing Television Broadcast Service: Fifth Report & Order*, 12 F.C.C. Rec. 12809, 12811-12812 (1997) (hereinafter FIFTH REPORT). In the Balanced Budget Act of 1997 (47 U.S.C. § 309(j)(14) (2004)), Congress codified the December 31, 2006 date.

[2] For more information on DTV, see CRS Report RL31260, *Digital Television: An Overview*, by Lennard Kruger.

[3] However, it should be noted that while duplication is fairly simple, distribution, especially of high-end digital content, can be quite time consuming. For example, over broadband connections, while a music file may take a matter of minutes, television shows in standard analog format take a number of hours. Digital programs (such as an hour of High Definition television programming) in turn, may take upwards of 10-15 hours to successfully download.

[4] FIFTH REPORT at 12811 ¶ 5.

[5] This collaboration was open to any group or individual wishing to participate, with the exception of the press. BPDG, *Final Report of the Co-Chairs of the Broadcast Protection Discussion Subgroup to the Copy Protection Technical Working Group*, FN 4, (June 3, 2002) (available at http://www.cptwg.org/Assets/TEXT%20FILES/BPDG/BPDG%20Report.DOC]).

[6] FCC, *In the Matter of Digital Broadcast Content Protection: Report and Order and Further Notice of Proposed Rulemaking*, MB Docket No. 02-230, 18 F.C.C.R. 23550, 23589 (November 4, 2003) (hereinafter REPORT AND ORDER).

[7] The amount of copy protection has a potential for variability. For instance, a content provider such as C-SPAN may decide that no copy protection is necessary, and would set the flag to an off-position. Digital content could therefore be available without any restrictions to the user. However, a broadcaster who sought to show pay-per-view digital content might choose to set the flag to an on-position, which would disallow any form of copying, and could also potentially add a setting to restrict the amount of time a user has to watch the program after purchase. Alternatively, a content provider may decide that individual copying is permitted, provided a user views that copy on a secure, compliant device.

[8] REPORT AND ORDER at 23561.

[9] REPORT AND ORDER, Appendix B, at 23592.

[10] 47 U.S.C. 336(i)(1) (2000).

[11] 47 U.S.C. 336(b)(4) and (5).

[12] 47 C.F.R. 73.624 (2004). *See also*: [http://www.fcc.gov/mb/policy/dtv/] for an overview of the FCC's activities with regard to the implementation of DTV.

[13] *Consumer Electronics Association v. FCC.*, 347 F.3d 291, 301 (D.C. Cir. Oct. 28, 2003).

[14] *U.S. v. Southwestern Cable Co.*, 392 U.S. 157, 178 (1968).

[15] As to ancillary jurisdiction, see *Southwestern Cable*, at 178. Concerning the FCC's ancillary authority, Congress has given the Commission "a comprehensive mandate," with "expansive powers," which has led the courts to conclude that the Communications Act of 1934 provides the Commission with ancillary jurisdiction over matters that are related to the provision of radio or television service, though not specifically enumerated in the act. Historically the FCC has exercised its ancillary jurisdiction to promulgate regulations in a number of areas. See *Southwestern Cable*, at 173, 177; *U.S. v. Midwest Video Corp*, 406 U.S. 649 (1972). In addition to this historic authority, the FCC relies on the definition of "wire/radio communications," which includes "all incidental 'instrumentalities, facilities, apparatus and services' that are

used for the 'receipt, forwarding, and delivery' of such transmissions" as a basis for its authority over television receivers. REPORT AND ORDER at 23563.

[16] FCC, REPORT AND ORDER, at 23575.

[17] *Id.* at 23555.

[18] *Id.*

[19] An owner of a copyright has a number of exclusive rights under the Copyright Act (17 U.S.C. § 101 *et seq.*), including the exclusive right to reproduce and distribute copies. To establish a claim of copyright infringement, the plaintiff must show ownership of the copyright and copying by the defendant. However, a claim of copyright infringement is subject to certain statutory exceptions, including the fair use exception (17 U.S.C. § 107 (2004)). This exception "permits courts to avoid rigid application of the copyright statute when, on occasion, it would stifle the very creativity which that statute is designed to foster" (*Dr. Seuss Enters., L.P. v. Penguin Books USA*, 109 F.3d 1394, 1399 (9th Cir. 1997)).

[20] Center for Democracy and Technology, *Implications of the Broadcast Flag: A Public Interest Primer*, (December 2003) (available at [http://www.cdt.org/copyright/031216broadcastflag.pdf]).

[21] REPORT AND ORDER at 23578.

[22] *Id.* at 23568-23569 (internal citation omitted).

[23] 2005 U.S. App. LEXIS 7847 (May 6, 2005).

[24] *Id.*

[25] *Id.* at 4.

In: New Intellectual Property Issues
Editor: John V. Luong, pp. 165-170

ISBN: 978-1-60021-211-6
© 2007 Nova Science Publishers, Inc.

Chapter 10

THE GOOGLE BOOK SEARCH PROJECT: IS ONLINE INDEXING A FAIR USE UNDER COPYRIGHT LAW?[1]

Robin Jeweler

SUMMARY

Google, Inc. is digitally scanning the collections of several prominent libraries in order to create a vast searchable database of literary works. Copyright holders who have not authorized and object to the digitization have filed suit against the company. This report provides background on the pending litigation. It will be updated as judicial developments warrant.

Complaints for copyright infringement were recently filed against Google, Inc. by a variety of authors and representatives of the book publishing industry.[1] The complaints specifically challenge Google's "Print Library" project, recently renamed "Google Book Search," an effort by Google in conjunction with several library partners to scan books into a digital format so that they may be searched textually. Although the case is in its very early stages, the issues presented have captured national attention. Once again, new technology and traditional principles of copyright law appear to be in conflict. Because of the unique facts and issues presented, there is scant legal precedent to legitimize Google's claim that its project is protected by copyright law's fair use exception to liability for infringement.[2] Thus, questions presented may be ones of first impression for the courts.

GOOGLE'S BOOK SEARCH PROJECT

In December of 2004, Google announced a partnership with several major libraries to make digital copies of their collections and permit the text of the literature to be searched online by the Google search engine.[3]

Google is providing its partnering libraries with a digital copy of the donor institution's collection.

The Print Library Project is only one of several initiatives by the company to enhance the breadth of its online search capabilities. It was originally a component of "Google Book Search," which included the Library Project and its "Partner Program," an online book

[1] Excerpted from CRS Report RS22356, Updated December 28, 2005

marketing program designed to help publishers and authors promote their books by displaying a limited number of sample pages in connection with a user's word search. The Library Project was described on its website:

> When you click on a search result for a book from the Library Project, you'll see the Snippet View which, like a card catalog, shows you information about the book plus a few snippets — a few sentences of your search term in context. You may also see the Sample Pages View if the publisher or author has given us permission or the Full Book View if the book is out of copyright. In all cases, you'll see 'Buy this Book' links that lead directly to online bookstores where you can buy the book.[4]

In addition, "Google Video" offers a search vehicle for material from archived television programs, educational videos, personal productions, and other video media. Users can search the closed captioning and text descriptions of its video archive for relevant results. However, in the case of both the Book Partner Program and the video program, content owners actively submit their material for inclusion in the searchable database. Only the Library Project does not seek authorization to copy from content owners. Hence, it is the only program that has been challenged in court.

After some academic and commercial publishers objected to the Library Project, Google took a brief hiatus from scanning to allow publishers time to identify works that they, i.e., the copyright holders, do not want to be included in the digital database. This has been referred to an "opt out" plan. The general rule of copyright law requires a prospective user to seek permission for use; Google has reversed the process by announcing its intention to digitize entire collections of the contributing libraries *unless* a content owner opts out by acting to *withhold* permission. This contributes to the content holders' claim that Google is engaged in massive copyright infringement.

The Parties' Positions

The complaint filed by plaintiff publishing companies (the Publishers) accuses defendant Google of massive copyright infringement. By digitizing copyrighted works without permission, Google is alleged to violate the copyright holders' exclusive rights to copy and/or display protected work.[5] Plaintiffs contend that Google's project is strictly commercial because it "pays" for the libraries' collections by delivering digital copies back to them; and, Google will realize significant advertising revenues as a consequence of its enhanced search capabilities.

Defendant Google essentially contends that its opt out program negates any infringement liability. But, if infringement were found, Google argues that its activity is protected by copyright's fair use doctrine. Google cites the U.S. Court of Appeals for the Ninth Circuit's decision in *Kelly v. Arriba Soft Corp.* as support for the proposition that Internet search engines' indexing activities constitute a fair use.[6]

THE LAW

As stated above, copyright law confers on the rights holder the exclusive right to control reproduction, display, and distribution of a protected work. Accordingly, in order to use a copyrighted work, one seeks permission from the owner and negotiates the terms, conditions, and payments for use. Google claims that for it to actively seek permission from every rights' holder in the multi-library collections would be impractical and prohibitive. The Publishers claim that Google's "opt out" program "stands copyright law on its head."[7] One cannot, they argue, generally announce one's intention to infringe multiple copyrighted works and collectively offer rights holders the opportunity *not* to have their work infringed.

Fair Use

Assuming a court were to find that Google's digitization of copyrighted works is infringing, the question becomes whether its activities are a fair use. The fair use exemption derives from common law and the First Amendment.[8] As codified in the Copyright Act, it establishes criteria for a court to consider in determining whether an infringing use is "fair." Specifically, 17 U.S.C. § 107 provides that, notwithstanding other provisions of the law, use of a copyrighted work

> ... for purposes such as criticism, comment, news reporting, teaching (including multiple copies for classroom use), scholarship, or research, is not an infringement of copyright. In determining whether the use made of a work in any particular case is a fair use the factors to be considered shall include —
> (1) the purpose and character of the use, including whether such use is of a commercial nature or is for nonprofit educational purposes;
> (2) the nature of the copyrighted work;
> (3) the amount and substantiality of the portion used in relation to the copyrighted work as a whole; and
> (4) the effect of the use upon the potential market for or value of the copyrighted work.

Because fair use is an "equitable rule of reason" to be applied in light of the overall purposes of the Copyright Act, other relevant factors may also be considered.[9]

The court hearing the case will make findings of fact and assign relative value and weight to the fair use factors in its analysis. This report will not attempt to predict an outcome in the pending litigation but does make some observations with respect to fair use analysis and the issues at hand.

The Library Project has the potential to be a great boon to scholarship, research, and the public in general. It is, nevertheless, commercial in nature because Google anticipates that it will enhance its service's utilization by the public and concomitantly increase advertising fees. With respect to the first factor, the purpose and character of use, the searching and indexing goal appears to be a highly transformative use of the copied text. There is little question that indexing basic information about any book alone, absent copying, would not constitute copyright infringement. While displaying "snippets" of text is closer to infringing activity, the prospective display, as described by Google, does not appear to usurp or negate the value of the underlying work.

The second factor is the nature of the copyrighted work. Digitizing the collections of the named libraries will encompass both factual and creative works, the latter being entitled to the highest level of copyright protection. How the court views the third factor — amount of the portion used — will be significant. In order to create its mega-database, Google will scan the entire copyrighted work, a major consideration weighing against fair use. But it intends to display, i.e., use, at any given time, only brief excerpts of the searchable text. Hence, is the digital reproduction incidental to an otherwise fair use or is it impermissibly infringing?

Finally, what will be the Library Project's effect on the potential market for or value of the copyrighted works? Here, Google makes a strong argument that its indexing and text searching capability has the potential to greatly enhance the market for sales for books that might otherwise be relegated to obscurity. Its "sampling" of text permits members of the public to determine whether they wish to acquire the book.

The Publishers counter that copyright owners routinely receive license fees for authorized sampling. Google's project may deprive them of the opportunity to participate in the creation of similar databases over which they have control and input. The Publishers have also expressed concern that the digital edition of the work Google returns to the participating library may facilitate piracy and/or additional unauthorized uses.[10]

Case Law

Although a court's finding that there is a fair use exception to copyright infringement is context-specific, it naturally looks to precedent for guidance. Google asserts that *Kelly v. Arriba Soft Corp.* supports its claim of fair use, and in many respects it does. In *Kelly*, the court found that Arriba Soft's search engine, which, in response to a user's inquiry, compiled a database of images by copying pictures from websites (without authorization) and displayed them as thumbnail images, was a sufficiently transformative use of copyrighted material to be a fair use. Although providing indexing information alone does not implicate copyright infringement, displaying limited quotes from a literary work may be consistent with fair use. Indeed, a quote from a literary work is a far more limited reproduction and display than a thumbnail image of a full-sized one. However, a major distinction between the *Kelly* case and Google's Library Project is that in the former, content owners voluntarily uploaded their images to the Internet.[11] Here, the search engine is copying material to create a new database to enable the search capability.

Hence, the arguably unique question presented is whether apparent *prima facie* infringing activity that facilitates an arguably legitimate use is indeed a fair one. In this broad respect, the Supreme Court's decision in *Sony Corporation of America v. Universal City Studios*[12] is apposite. In *Sony*, the Court held that the sale of the video recording machine to "time shift" broadcast television for personal home viewing was not contributory copyright infringement. The Court articulated a new category of fair use, namely, time shifting. Although the factual underpinnings and legal precedent of this decision are *not* particularly relevant to nor controlling in the instant case, the *Sony* decision itself stands as a landmark in copyright law demonstrating the willingness of the Court to balance new technological capablilites against traditional principles of copyright law.

Although *Sony* sanctioned "time shifting" of in-home television broadcasting, neither the U.S. Supreme Court nor the lower courts have evidenced willingness to expand this judicially created category of fair use. In *UMG Recordings v. MP3.Com, Inc.,* a U.S. district court rejected out-of-hand the defendant's proffered fair use defense as a justification for unauthorized copying of plaintiffs' audio CDs.[13] The defendant claimed that its unauthorized copying enabled it to provide a service to proven owners of a CD, namely "space shifting," or enabling them to obtain access to the music on the CD through its subscription service.

In the case of Google, many copyright experts see analogies to the technological considerations inherent in *Sony*. The digital scanning — the alleged infringing activity — is viewed as incidental to a valid and socially useful function, indexing.[14] Others view Google's activity as *prima facie* copyright infringement, with little or no extenuating circumstances.[15]

CONCLUSION

How the court (or courts) that consider this case define the issues presented will ultimately determine whether the suit against Google sets an important precedent in copyright law. Viewed expansively, the court may find that copying to promote online searching and indexing of literary works is a fair use. To many observers, such a holding could be the jurisprudential equivalent of *Sony's* sanctioning of "time shifting." If the court adopts a more narrow view of fair use that precludes Google's digitization project, searchable literary databases are likely to evolve in a less comprehensive manner but with the input and control of rights holders who view them as desirable and participate accordingly.

REFERENCES

[1] McGraw-Hill Companies, Inc. v. Google, No. 05 CV 8881 (S.D.N.Y., filed Oct. 19, 2005); The Author's Guild v. Google, No. 05 CV 8136 (S.D.N.Y., filed Sept. 20, 2005).

[2] 17 U.S.C. § 107.

[3] Participating libraries include those at the University of Michigan, Harvard University, Stanford University, Oxford, and the New York Public Library. For details, *see* [http://books.google.com/ googlebooks/partners.html].

[4] [http://books.google.com/googlebooks/library.html]. For more background on the program, see Google's website at [http://books.google.com/googlebooks/about.html].

[5] 17 U.S.C. § 106.

[6] 336 F.3d 811 (9th Cir. 2003). For more background on Kelly v. Arriba Soft Corp., see CRS Report RS21206, *Fair Use on the Internet: Copyright's Reproduction and Public Display Rights,* by Robin Jeweler.

[7] *Publishers: Value of Book Search Project Shows That Scanning Is Not Fair Use,* 71 BNA PATENT, TRADEMARK & COPYRIGHT J. 94 (Nov. 25. 2005).

[8] Eldred v. Ashcroft, 537 U.S. 186, 219 (2003).

[9] Sony Corp. of America v. Universal City Studios, 464 U.S. 417, 448, 454 (1984).

[10] In addition to fair use, the Copyright Act contains additional limitations on exclusive rights of copyright holders. 17 U.S.C. § 108, for example, gives certain reproduction

rights to libraries and archives, but these provisions are not relevant to the issue discussed.

[11] Some observers suggest that uploading to the Internet may be construed as a limited license for foreseeable uses such as those by Internet search engines. Jonathan Band, *The Google Print Library Project: Fair or Foul?,* 9 J. OF INTERNET LAW 1, 4 (Oct. 2005).

[12] 464 U.S. 417 (1984).

[13] 92 F.Supp.2d 349, 352 (S.D.N.Y. 2000)("[D]efendant's 'fair use' defense is indefensible and must be denied as a matter of law.").

[14] *See, e.g.,* Jonathan Band, *supra;* Christopher Heun, *Courts Unlikely to Stop Google Book Copying,* INTERNETWEEK (Sept. 2, 2005) at [http://internetweek.cmp.com/showArticle.jhtml? articleID=170700329].

[15] *See, e.g.,* Elisabeth Hanratty, *Google Library: Beyond Fair Use?,* 2005 DUKE L. & TECH. REV. 10 (2005); Raymond Nimmer, *Google Lawsuit Begins; Fair Use* (Oct. 3, 2005), online at [http://www.ipinfoblog.com/archives/intellectual-property-33-google-lawsuit-begins-fair-use. html].

INDEX

D

E

F

Q

R

S

T

U

V

W